Sound Structure in Music

Robert Erickson

SOUND STRUCTURE IN MUSIC

University of California Press
Berkeley   Los Angeles   London
1975

University of California Press
Berkeley and Los Angeles, California
University of California Press, Ltd.
London, England
Copyright © 1975 by The Regents of the University of California
ISBN: 0-520-02376-5
Library of Congress Catalog Number: 72-93520
Printed in the United States of America

For Lenore

Contents

Preface

This book was written for musicians, especially for composers, with more than a glance in the direction of performers of contemporary music. It aims at gaining some understanding of the role of timbre in music and at locating principles upon which a theory of timbre organization might be constructed.

I have approached other disciplines as a musician, looking for facts, ideas, points of view that might help us to view our art from other perspectives, avoiding technical jargons whenever I could. I hope I have not misconstrued or oversimplified concepts taken from physical acoustics and experimental psychology by adapting them to specifically musical situations; I have tried not to reduce the richness of music to conceptual schemata.

I wanted to look closely at how we have used timbre in music, at our assumptions about its functions, and at our traditional ideas about what we are doing. I hope I have offered no recipes for constructing timbral music. That needs ears and imagination above all. This is more a what-might-be than a how-to-do-it book, for all its emphasis upon practice.

Sound Structure in Music

1

The Sounds Around Us

We cannot close our ears; we have no ear lids. When a sound is loud enough, or especially meaningful—one's name, a baby's cry—one may react to it though asleep. Sounds that are unimportant to us are ignored, asleep or awake, even when quite loud. Not that we do not hear them: we are always aware of the whole passing sound stream while selecting some sounds for focal attention.

When we hear a strange sound—thinking "What was that?"—we focus upon it, listen carefully, wait for a repetition. We try to identify it. Once it has been identified, recognized, we relax. It has been related to other sounds we know. More particularly we have identified whatever object or process or animal or person caused or emitted the sound. The usual sounds of daily life are often cues to actions: think of a telephone bell or an automobile horn. Occasionally we may pay attention to the sound itself. Then it is more than a cue, and we are listening in another mode, music mode, regardless of the source of the sound. Many people listen to the sounds of birds and crickets in this manner. The whole of our sound world is available for music-mode listening, but it probably takes a catholic taste and a well-developed interest to find jewels in the auditory garbage of machinery, jet planes, traffic, and other mechanical chatter that constitute our sound environment. Some of us, and I confess I am one, strongly resist the thought that it is garbage. The more one listens the more one finds that it is all jewels.

THE SOUNDS OF SPEECH

During the past few decades much scientific research has gone into the study of speech and speech sounds. Investigators have found evidence that people process speech differently than other sounds. These findings have led to the startling suggestion that we may be "wired up" for speech, that the capacity for speech is innate in humans.

Eric. H. Lenneberg has pointed out certain universals: all languages use phonematization; all languages concatenate words into sequences (phrase, sentence, discourse); all languages have some sort of grammar; a child "makes progressive use (in the sense this word is employed in genetic theory) of environmental stimulating or releasing mechanism, not because the child's "long range purpose is to speak like the neighbor," but because he is endowed with a complex series of innate, interacting behaviour patterns that are elicited in a more or less automatic manner (i.e., unplanned by parents) by the speaking environment that surrounds him (1960:891)."

Evidence has been offered that one ear enjoys an advantage in speech perception, that "while the general auditory system may be equipped to extract the auditory parameters of a speech signal, the dominant hemisphere is specialized for the extraction of linguistic features from these parameters" (Studdert-Kennedy and Shankweiler, 1970:592).

The probability that there is a special sort of sound processing identified with the speech mode should make us wary about drawing musical inferences from research into speech perception. On the other hand speech sounds have been an integral part of song for thousands of years. Most of the songs we have sung have had words to them. All of them have had vowels. Furthermore, if the scientific community eventually accepts evidence that our capacity for speech is innate, then most of this evidence will also apply to song, if not to all music. At the very least we can use some of the insights of speech research, especially the work on speech sounds, to gain some understanding of musical sounds and musical quality or timbre. Speech and music both depend upon contrasts between sounds. The basic units for speech are phonemes and the basic units for music are timbres or simply sounds.

Detailed study of the sounds of speech in modern times has led investigators away from a narrow concern with the physical description of speech sounds. G. A. Miller pointed out,

In order to discuss even the simplest problems in speech production and speech perception, it is necessary to be able to distinguish significant from nonsignificant aspects of speech. And there is no simple way to draw this distinction in terms of the physical parameters of the speech signal itself. Almost immediately, therefore, we are forced to consider aspects of language that extend beyond the acoustic or physiological properties of speech, that is to say, beyond the objective properties of "the stimulus." [1965:204]

P. B. Denes goes into more detail:

The basic premise of (most speech-recognition) work has always been that a one-to-one relationship existed between the acoustic event and the phoneme. Although it was recognized that the sound waves associated with the same phoneme could change according to circumstances, there was a deep-seated belief that if only the right way of examining the acoustic signal was found, then the much sought-after one-to-one relationship would come to light. Only more recently has there been a wider acceptance of the view that these one-to-one relationships do not exist at all: the speaker produces acoustic signals whose characteristics are of course a function of the phoneme to be currently transmitted, but which are also greatly affected by a variety of other factors such as the individual articulatory characteristics of the speaker, the phonetic environment of the sound to be produced, linguistic relationships, etc. As a result, the acoustic characteristics of the sound to be produced do not identify a particular phoneme uniquely, and the listener resolves the ambiguities of the acoustic signal by making use of his own knowledge of the various linguistic and contextual constraints mentioned above. The large part that is played by these nonacoustic factors in the recognition of speech is best shown by the remarkably small loss of intelligibility produced by quite serious distortions of the speech wave. [1964:892]

Thus, understanding of speech sounds must be within speech, and, although we may not know a great deal about how it is processed in the human auditory system, we may expect that processing to be very different from a passive representation of acoustic events.

Objections may be raised that musical timbre should be equated with the quality of the speaking voice rather than with the individual speech sounds; that it is the "clarinet quality" rather than the individual sounds which is significant, and that the overall "clarinet quality" corresponds to the quality of a speaker's voice. I do not mean to exclude this aspect of timbre. But overall "clarinet quality" can be shown to have no clear-cut one-to-one relationship to the acoustical signal either! We can no more synthesize a clarinet from a single physical description of the signal than we can synthesize all the *ah* sounds we use in speech from a single acoustical recipe. Analysis of the gamut of clarinet tones might lead one to say that it is three instruments, rather than one! We are back to the problem of individual sounds.

Another objection might be that individual speech sounds and individual musical timbres have never been shown to have anything in common, and that until this has been demonstrated one should not look upon speech sounds as a class of timbres.

A few years ago A. W. Slawson (1965, 1968) performed some pertinent experiments. He synthesized vowel-like sounds from averages derived from the acoustical analysis of many spoken vowels. Sometimes he told his listeners to judge differences in terms of vowel quality. Other listeners were asked to judge the same sounds in terms of instrumental color or timbre. His results suggested "that vowel quality and musical timbre are similar functions of their acoustic correlates in a large class of sounds to which both concepts can be applied" (1968:101).

It is a long way from Slawson's synthesized stimuli to the sounds of musical instruments and the natural speech of real persons, but one may reasonably expect psycholinguistics and psychoacoustics to offer insights about the mental processing of speech which could help us to understand processing of timbre in music.

If it is true that the understanding of the sounds of speech must be within speech, then the same statement should be true for music. As a consequence of our very recently expanded electronic and computer technology, our accepted working concepts and rules of the thumb are badly in need of clarification. These clarifications must be relevant to musical situations. We cannot expect physics, linguistics, or psychology to provide ready-made theories for us, however much their discoveries may help. Any theory of musical timbre will have to build itself from music.

TIMBRE, WHAT IS IT?

If we are to learn more about what we are doing we are first of all going to need a better definition of musical timbre, and, alas, the definitions usually used by musicians are murky or contradictory, or both.

The American Standards Association publishes recommended definitions in many fields, and their 1960 (the latest at this writing) definition of timbre is typical of many in suggesting that timbre is that attribute of sensation in terms of which a listener can judge that two sounds having the same loudness and pitch are dissimilar. In other words, everything about a sound which is neither loudness nor pitch might be timbre. A note to the definition adds: "timbre depends primarily upon the spectrum of the stimulus, but it also depends upon the waveform, the sound pressure, the frequency location of the spectrum, and the temporal characteristics of the stimulus" (1960:45).

Clearly timbre is a multidimensional stimulus: it cannot be correlated with any single physical dimension.

Until fairly recently most researchers have studied the spectral dimension of timbre without paying much attention to those matters mentioned in the note appended to the ASA definition, because changes in spectrum do produce differences in timbre, and because much important work remains to be done in this area. Nevertheless, dissatisfaction has been expressed, for example, by R. W. Young:

In discussions of timbre (tone quality) it has long been the custom to state that differences in quality of tone are solely dependent on the occurrence and strength of partial tones. Although H. von Helmholtz, in making this statement, recognized that the characteristic tone of some instruments is dependent upon the way the tone stops and starts, he chose to restrict his attention to the "peculiarities of the musical tone which continues uniformly" and to consider as musical only those tones with harmonic upper partials. Many writers since have adopted these simplifying but not realistic assumptions; according to such simplifications the piano is not a musical instrument!

The transient parts of a sound contain important clues by which different instruments are identified. A sustained high tone on the clarinet, for example, is practically indistinguishable from the same tone (sustained) played on the flute, but the initiation of the sound is likely to be noticeably different on the two instruments.

Another distinctive characteristic of the tone quality of an instrument is the formant, or frequency range within which the partials of the sounds emitted by the instrument have relatively large amplitudes. [1960:661]

That a physicist should express such strong complaints about simplifying assumptions is a sign that the definition has been so narrowed that it is almost irrelevant. In attempting to get at the most important dimensions the essential has disappeared.

Descriptions of the physical correlates of timbre which try to avoid simplifying assumptions may be more difficult to quantify but they are likely to have more to say to a musician. The short exposition by James C. Tenney (1965) takes into account the transient phenomena mentioned by Young, and includes modulation processes (vibrato, tremolo, and others) characteristic of sounds produced by musical instruments. Using computer synthesis, where the simulus could be clearly specified, and himself as the listening judge, he came to the conclusion that, although spectrum, transient phenomena and quasi steady-state modulation processes may be the most important dimensions, each of these is characterized by a great many subparameters, and that the definitions based upon Ohm's Law (see chapter 2) are inadequate for any definition of timbre which might be of musical value.

The trouble with a definition as spacious as Tenney's is that one immediately has to ask whether some of the subparameters could not safely be left out of one's calculations. Anyone making a computer program wants it to be as simple as possible: computer time is expensive, and perhaps certain nuances and refinements cost too much for their musical value. Before we can assess the importance of a multitude of microdimensions we need an overview that describes the most important aspects of timbre and puts the whole problem into some sort of perspective. Recently J. F. Schouten offered a set of dimensions for timbre which are scaled to the concerns of much contemporary music:

Summing up, the elusive attributes of timbre can be considered to be determined by at least five major acoustic parameters:
1. The range between tonal and noiselike character.
2. The spectral envelope.
3. The time envelope in terms of rise, duration and decay.
4. The changes both of spectral envelope (formant-glide) and fundamental frequency (micro-intonation).
5. The prefix, an onset of a sound quite dissimilar to the ensuing lasting vibration. [1968:42]

Here one has a set of music-oriented concepts for thinking about timbre, whether noises, pitches, vocal sounds, traditional instrumental sounds, electronic, or any other sounds. Moreover, for each of these dimensions subjective experience and physical phenomena can be related.

SUBJECTIVE	*OBJECTIVE*
Tonal character, usually pitched	Periodic sound
Noisy, with or without some tonal character, including rustle noise	Noise, including random pulses characterized by the rustle time (the mean interval between pulses)
Coloration	Spectral envelope
Beginning/ending	Physical rise and decay time
Coloration glide or formant glide	Change of spectral envelope
Microintonation	Small change (one up and down) in frequency
Vibrato	Frequency modulation (or amplitude modulation)
Attack	Prefix
Final sound	Suffix

The five dimensions are fundamental to any discussion of timbre where subjective and objective phenomena are to be correlated, and they constitute an excellent set of categories for the kinds of perceptual analysis which seem important for music. Tenney, Young, and the writers of the ASA definition touched upon most of these parameters. The only new idea here, and it is a very important one for music, is that of rustle noise. In acoustics noises are usually characterized as to the spectral envelope: white noise is defined as having a constant spectral density in whatever frequency range is being considered; if the spectral density is not constant the noise is colored. The individual pulses are very closely spaced, and, according to Schouten, if the mean distance between pulses is 0.3 milliseconds or less rustle noise will be indistinguishable from white noise proper. The various noises of nature and musical instruments present a rather wide variety of pulse spacings, and that is why the concept of rustle time, which is measurable and which can be correlated with physical properties of the sound, is so valuable.

TIMBRE OR TONE COLOR?

The word or phrase we use as a handy reference is much less important than the idea to which it refers. I generally use the term "timbre" but often use "tone color" in the same sense. Neither term is very satisfactory, nor is "tone quality" much of an improvement. The English language has nothing better to offer, however, as Alexander Ellis, the translator of Hermann Helmholtz's *On the Sensations of Tone*, discovered to his dismay. His remarks are worth presenting at length:

Prof. Helmholtz uses the word *Klang* for a *musical tone,* which generally, but not always, means a *compound tone.* Prof. Tyndall therefore proposes to use the English word *clang* in the same sense. But clang has already a meaning in English, thus defined by Webster: "a sharp shrill sound, made by striking together metallic substances, or sonorous bodies, as the *clang* of arms, or any like sound, as the *clang* of trumpets. This word implies a degree of harshness in the sound, or more harshness than *clink.*" Interpreted scientifically, then, *clang* according to this definition, is either *noise* or one of those *musical tones with inharmonic upper partials* which will be subsequently explained. It is therefore totally unadapted to represent a *musical tone* in general, for which the simple word *tone* seems eminently suited, being of course originally the tone produced by a *stretched* string. The common word *note,* properly the mark by which a musical tone is written, will also, in accordance with the general practice of musicians, be used for a *musical tone,* which is generally compound, without necessarily implying that it is one of the few recognised tones in our musical scale. Of course, if *clang* could not be used, Prof. Tyndall's suggestion to translate Prof. Helmholtz's *Klangfarbe* by *clangtint* fell to the ground. I can find no valid reason for supplanting the time-honoured expression *quality of tone.* Prof. Tyndall quotes Dr. Young to the effect that "this quality of sound is sometimes called its register, colour, or timbre." *Register* has a distinct meaning in vocal music which must not be disturbed. *Timbre,* properly a kettledrum, then a helmet, then the coat of arms surmounted with a helmet, then the official stamp bearing that coat of arms (now used in France for a postage label), and then the mark which declared a thing to be what it pretends to be, Burn's "Guinea's stamp", is a foreign word, often odiously mispronounced, and not worth preserving. *Colour* I have never met with as applied to music, except at most as a passing metaphorical expression. But the difference of tones in *quality* is familiar to our language. . . . [Helmholtz, 1877, trans. Ellis, 1954:174]

Ellis lost his fight to discredit that foreign word, timbre. It has been Anglicized to the point where I have occasionally seen it spelled timber. A people who could invent a pronunciation for ancient Greek comfortable for English throats would hardly mind defrenchifying a word. Tone quality is used today, but there is no denying it is a clumsy phrase. Too bad that *clangtint* did not get established. By now it could have been well worn with usage and rich in associations.

A UNIVERSE OF TIMBRES! DREAMS AND DISAPPOINTMENTS

When it became clear, in the early fifties, that musically useful sounds could be created from electronic signal generators, composers began to dream of a future unlimited by the sounds of conventional instruments. Henri Pousseur tells of some expectations at the Cologne studio:

Stockhausen felt that . . . if one could consider a complex sound, a complex wave, as the *sum* of simple components, then one could also produce such a complex sound, with absolute precision of control, by putting together, by mixing these different components. And so, using only sine waves, and through their assemblage, one would be able to rebuild, or to build originally, any imaginable sound event. . . . In these first attempts, we were still very far from the desired goal. Instead of a situation in which the sine tones came together to form more complex sounds, they remained basically discrete and identifiable; we had a situation in which the sine wave material was used like an easily recognizable instrument. [1968:22]

What a disappointment this must have been! All that effort and no universe; only the sounds of sine tone mixtures, "sometimes (with a decrease in volume) like a very sweet, attackless vibraharp, sometimes (with more sustained sounds) like the softest tones of a pipe organ" (1968:22). The difficulty was that precise control of frequencies, attacks and decays, amplitudes of individual components, and noise content could not be attained at the Cologne studio.

Simulation of certain conventional instrument sounds has now been successfully accomplished, and vast amounts of electronic music have been composed in the past twenty years. Today we are aware that the synthesis of a single acceptable instrument sound can be a difficult, time-consuming, and expensive undertaking, and that "interesting" sounds are likely to have a very detailed microstructure. And we are discovering a rather depressing uniformity among compositions made on the same electronic generating equipment. Improved technology may remedy the purely technical difficulties, but some important musical problems have been uncovered too. For it appears that there are fewer musically different electronically generated sounds than one would have expected. Listeners tend to hear electronic sounds as a class—"ah, electronic"—just as they might hear "ah, a violin." This is discouraging, because these are the same listeners who are able to make distinctions such as "orchestra," "strings," "brass," "violins," "trombones," "cellos," "trumpets"; and "flute," "clarinet," "oboe," "bassoon," "French horn," "tuba."

Learning? Perhaps, but there is more to it than that. Now that we are able (at least in principle) to produce any imaginable sound we are going to have to think about which ones are worth making for any particular composition. How shall we make our decisions?

Basic distinctions in electronic music are between noises and periodic sounds. The noises may be filtered in an infinity of ways. The periodic sounds, sine, square, pulse, may be combined, and the complex periodic sounds filtered, in an infinity of ways. An attack generator can supply a variety of on-time speeds and amplitudes.

A composer who wishes to carve out certain sounds from this infinity of possibilities must decide: which ones? He may attempt to create "an instrument," meaning some sort of unified selection of sounds from the infinity of possibilities. This is an avenue that has been explored by some of those composing computer music, and the results have been somewhat disappointing. Or he may go at things more abstractly, thinking in terms of contrast, similarity, sound classes. He will discover the paradox that the more he tries for an infinity of timbres the more he will tend toward no significant contrasts. He may accept this situation or even make it the basis of his music, eliminating the possibility of any significant organization of timbre. For those composers who try to organize a large variety of contrasting sounds the puzzle is still there.

We have the same problem in music with regard to pitch. Pitch may be a continuum but the pitches of music have always been concerned with discrete intervals and their nuances. Are discrete timbres—"that's a bell"; "that's a whosis"—in any way similar? It may be true that we are on the edge of being able to produce any sound we can imagine, just as it is true that we can produce any pitch we can imagine. The infinity of sounds in the universe of timbres may be objectively real to physics and measuring instruments; if it is unrealizable in music then the difficulty must be related to human limitations and to the limitations imposed by musical discourse.

RECOGNIZING AND IDENTIFYING

The most striking thing about our subjective sound experience is that we hear multidimensional sounds as unified perceptual objects. Even if we try very hard we find it difficult to attend to any single parameter of a timbre. Schouten puts it this way: "Evidently our auditory system does carry out an extremely subtle and multi-varied analysis of these elements, but our perception is cued to the resulting overall pattern. Acute observers may bring some of these elements to conscious perception, like intonation patterns, onsets, harshness, etc., even so, minute differences may remain unobservable in terms of their auditory quality and yet be highly distinctive in terms of recognizing one out of a multitude of potential sound sources" (1968:90).

The key word here is "recognizing." It is as essential in music as in other human activities. We recognize all sorts of things—musical intervals, scale patterns, harmonic complexes, phrases, sections, complete compositions. And many different musical instruments. Anyone can recognize familiar instruments, even without conscious thought, and people are able to do it with much less effort than they require for recognizing intervals, harmonies, or scales. We are able to follow a melody on a particular instrument, or several melodic lines on several different instruments, even when embedded in rather dense textures. There are limits: the texture must not be too dense; the tempo must not be too fast; the instruments which we are recognizing must be sufficiently contrasted—one could compile quite a long list of not too much/not too little.

When we recognize or identify a sound we are in fact classifying it. To recognize that a sound was made by a violin is to assign it to a class. And classifying means making categories. Paul Kolers points out, "the human visual system categorizes; it does not analyse" (1968:54), and it is just as true for the human auditory system. There is nothing absolute about the type or number of categories, and the way in which we categorize depends upon the situation, our background, training, and so on. Kolers says: "Once his input threshold is reached, the human will make something of the information. Unlike the computer, the human program rarely fails to run. Although he may not be efficient, *the human will do the best he can with what is presented to him, within the limits defined by the input and his modes of categorizing*" (ibid.:56; italics mine). The italicized phrase is not a bad description of the stance of a listener, a performer, or, for that matter, a composer.

Let me give a few instances, closer to musical life. My friend and I are sitting in an auditorium, row *J*, and someone is playing a musical instrument on stage. "What do you hear?" "Music." "Can you identify the instrument?" "Sure, it's a violin." "How do you know?" "Anyone can *see* it's a violin!" Short circuit. Not uncommon.

Another friend, same place. This one is proud of his ability to recognize instruments. "I tell you what, I'll close my eyes. Have him play just one single note and I'll guess what the instrument is." The player bows an open A string on his viola and my friend says, "Violin." Close, but no cigar. An easy confusion, given the circumstances. Many experienced musicians could be fooled.

I try a similar game on myself, the violin again. The player has promised to improvise, working in as many different kinds of sounds as he can. My task is to guess how he made them. The sounds fall into a number of categories for me: bowed sounds; plucked sounds; harmonics; *sul tasto* and *sul ponticello* bowing; even a few specialties, such as plucking with the left hand while simultaneously starting a bow stroke. Having once played the violin I not only knew these sounds; I had names for all of them and knew how they were produced. All this is hardly surprising—it was another sound he made, and he made it three times. The first time I thought he had scraped his foot on the floor accidentally; the second time I paid more attention; the third time it was clearly no accident and it had found a category, because when he finished his demonstration I asked, "What was that crunch?"

Each of the three listeners did the best he could with what he was presented, within the limits defined by the input and by his modes of categorizing in that player-in-auditorium situation. And we cannot leave out the situation, because it includes all sorts of things, such as the general musical expectations of the listener, the acoustical effects of the hall itself, including a (roughly) defined distance between player and listener and what, for now, we can call the "ambience" of the sound.

G. A. Miller has pointed out (1962) that our perceptual equipment is designed to make many measurements simultaneously, but to do a fairly crude job on each one. In extenuation we do have some ability to trade breadth for accuracy, this trading relation being part of the subjective phenomenon, attention. If we go back to our imaginary hall again, this time to listen to an orchestra, we shall see that Miller's observation describes a very complex set of mental operations.

Notice what we are able to do: we can listen to the orchestral sound as a whole; we can focus attention on one or another of the instrumental groups; occasionally we can even focus on a single instrument, isolating it from the mass of sound. The broader the focus the less the detail. We change from one to another category so easily that we may hardly be aware of how active this mental processing is; nor do we usually realize how much our modes of categorizing have to do with what we perceive.

SUBJECTIVE CONSTANCY AND TIMBRE

I mentioned earlier that an acoustical analysis of the sounds of a clarinet might lead one to expect two or three instruments rather than one. Nevertheless, under many circumstances we hear it as a single instrument. The approach to timbre from acoustics searches for invariants, taking the view that if we are able to recognize and identify a clarinet under conditions of changing pitch and loudness, in different environments and with different players, then as David Luce says, "The implication is that certain strong regularities in the acoustic waveform of the above instruments must exist which are invariant with respect to the above variables" (1963:17).

There certainly are some regularities, but they do not appear sufficient to completely account for our powers of recognition and identification. One needs some explanation of why we are able, for example, to follow a clarinet throughout its range, loud and soft, when there are so many irregularities in the acoustic waveforms of its various pitches.

Students of vision and visual perception have shown that colors appear to stay constant under changing conditions of illumination. K. Lorenz says that the subjective constancy of color "is the achievement of a very complicated 'calculation' by an unconsciously working apparatus within our central nervous system" (1961:171) and relates these "calculations" to the "unconscious conclusions" discussed by Helmholtz in his *Physiological Optics*. K. M. Sayre describes the "calculation" in terms of information processing:

The facts behind color-constancy phenomena, however, are that we require fine color discriminations less frequently than gross discriminations, and when gross discriminations enable us to maintain focus on objects of prime interest, we "systematically overlook" differences beyond the necessary degree of fineness. The mechanism which accomplishes this "systematic overlooking" is the information-processing system of the organism, and the principle according to which it is accomplished is that this system never expends more of its capacity on a given perceptual task than is necessary according to the current needs and interests of the agent. [1968:151–152]

I believe that our ability to "follow" the sound of, say, a clarinet playing a melody can best be explained as an instance of subjective constancy, and that the mental processing is of the type discussed by Lorenz and Sayre.

TIMBRE AS CARRIER

Most traditional music makes use of this subjective constancy of timbre. Melodies do not switch instruments from note to note in the historical styles of Western music; the instrument is kept the same and the pitches change. The many measurable differences between timbres are "overlooked"; they are nuances. Furthermore, all other musical dimensions combine to enhance the subjective constancy of the timbre: much of the melodic motion is stepwise; skips will be few and structured in such a way that the timbre constancy does not "break apart." The player will have been schooled for many years to maintain timbre constancy in many musical situations. In the case of the clarinet he will attempt to maintain it even when making musical contrasts between the chalumeau and other registers, playing down registral contrasts, natural to the clarinet, to get more continuity through all registers.

The chief function of timbre in most Western concert music of the past has been that of carrier of melodic functions. The differences of timbre at different pitches and in different registers of instruments, and the different timbres produced by the voice singing different vowels (not to mention timbral differences in vocal registers) have been treated as nuances. The nuances have always been an extremely important source of those "irregular regularities" without which there could be no art at all; but in the fundamental structure of this music timbre has functioned as carrier.

TIMBRES AS OBJECTS

The sounds of a clarinet may also be used in such a way that their differences are thrown into relief, and the collection of sounds may be composed in such a way that the pitch/melodic part of the music is less important (less in the foreground) than timbre contrasts in the sound sequence. Gross changes of pitch register, dynamics or articulation, and the avoidance of stepwise motion and rhythmic regularity enhance the perception of a sequence of contrasting sound objects. This idea of a musical instrument as a collection of contrasting sounds rather than a source of sounds that are invariant under pitch change grows out of the interest of musicians in *klangfarbenmelodie* and musics where the "whole sound" participates.

Some sixty years ago Schoenberg coined the term and speculated about the musical possibilities of "melodies" comprised of timbres:

> If it is possible to make compositional structures from sounds which differ according to pitch, structures which we call melodies, sequences producing an effect similar to thought, then it must also be possible to create such sequences from the timbres of that other dimension from what we normally and simply call timbre. Such sequences would work with an inherent logic, equivalent to the kind of logic which is effective in the melodies based on pitch. All this seems a fantasy of the future, which it probably is. Yet I am firmly convinced that it can be realised. [1911:470–471]

With the advent of musique concrète and electronic and computer sound generation, the technical means for creating controlled timbral sequences has led composers toward musics that depend more upon contrast than upon constancy. Instruments are changing and instrumental means and techniques are expanding very quickly to meet the demands of those composers who are working with klangfarbenmelodie ideas.

KLANGFARBENMELODIE: CONTRAST AND CONTINUITY

The musical implications of linear timbre contrast are discussed at length in a later chapter, but I wish to offer a few examples here to show some differences between the ideas of timbre as carrier and timbre as object.

Consider the subject of Bach's six-voice *Ricercar* from the Musical Offering. The composition was written in open score, and no particular instrumentation was specified (fig. 1). The *Ricercar* can be performed, with difficulty, on a keyboard instrument; or it can be arranged for an ensemble of instruments.

FIGURE I

For Bach, and for most arrangers of this work, it was taken for granted that the subject would be performed in a single "homogeneous" tone color, such as an organ stop or a particular instrument. In other words, timbre would be treated as invariant, and it would function as carrier of the melodic motion.

In 1934–1935 Webern made an arrangement of this *Ricercar* for small orchestra where the subject is distributed among several instruments, the effect of which is related to the more radical klangfarbenmelodie of Schoenberg's description.

FIGURE 2

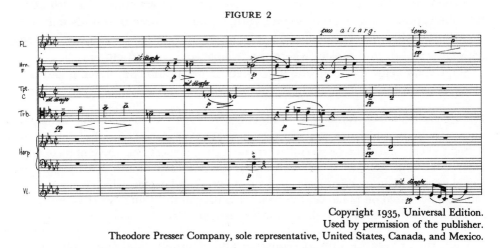

In figure 2, a passage from the opening measures of the *Ricercar*, the subject passes from trombone to horn to trumpet to horn to trombone to horn to trumpet. Measure 5 has an important accent in the harp, and the harp doubles the trumpet in measure 8.

Now examine the segments more carefully. The theme has been broken up in such a way that its joints, its articulations, are thrown into relief, and the segments are not all of the same length. Obviously, the use of timbre must be related in some way to the rhythmic organization of the phrase segments. What did Webern have in mind?

We need not guess, because in 1938 he wrote a fascinating letter to Hermann Scherchen about a performance for the BBC in which he revealed the reasoning behind his segmentation plan (Kolneder, 1961:154–155).

He explains that he thought of the theme as falling into two parts at the tenth note, the E♭ of the French horn in measure 5. Note the accent in the horn part, and the support for that accent from the harp. This E♭ is at once the final note of the first half of the theme and the first note of the second half (which also has ten notes.)

Webern breaks the first half of the subject into two segments of five notes each: five, undivided (trombone); four plus one (two horn, two trumpet, one horn). The second half is also broken into two five-note groups that are subdivided. It begins with one plus four (horn, trombone) and concludes with three plus two (horn, trumpet/harp). In measure 6 there is an overlap of the D; it is played by both horn and trombone, which tends to obscure the one plus four segmentation.

Thus the nineteen different notes of the theme are organized into two ten-note segments. Each ten-note segment is divided into two five-note groups. All of the five-note groups except the first are further divided as two plus two plus one; one plus four (which can also be read as two plus three in measures 5 and 6); and three plus two. This is no surprise, knowing what we do about Webern's propensity for symmetrical relationships of all sorts.

What is striking is that these segmentation patterns are counterpointed against the group meter. Against a duple background they superimpose patterns of fives summing to two large phrase segments of ten. The parameter is partly duration in a rhythmic sense, but it is not merely duration that is involved; it is something new—structured timbre. This structured timbre of the segment lengths is strong enough to have important effects upon other musical dimensions.

Notwithstanding the detailed timbral structuring that Webern brought to the composition of "his Bach Fugue," the full development of klangfarbenmelodie is constrained by the original Bach *Ricercar*. For more radical expressions of the melodic organization of timbres as objects we need to examine other works.

Webern's *Five Pieces for Chamber Orchestra*, Opus 10, is full of klangfarbenmelodie effects; for example, the final measure of the first piece, where a single pitch is distributed among several instruments (fig. 3).

The F above middle C goes from flute to trumpet/flute to trumpet to celeste. There is no sense of dislocation; rather, the effect is of timbre transitions by discrete steps. The most important transition is effected by the flute/trumpet doubling of the second F, acting as a link between flute alone and trumpet alone. At pianissimo levels and with the trumpet muted there is enough relationship between the two instruments for perceptual confusion, and Webern takes advantage of that fact. He is not only composing the contrasts between instrumental sounds, he is composing the range of similarities as well.

This concern for both continuity and contrast on the microlevel can be heard in the opening measures of the same piece (fig. 4). Each of the first three sounds is differentiated from the others: (1), harp and trumpet; (2), harp, viola harmonic, celeste; (3), harp and flute. This should make for considerable contrast, and it does. Nevertheless, notice that the harp is common to all three sounds, even though its second note is a harmonic. Also notice that the previously mentioned muted trumpet/flute relationship is involved and that there are obvious similarities between the viola harmonic and the flute at

FIGURE 3

pianissimo. In sum, it is clear that for Webern not just any timbre contrast will do. He is actually composing the contrasts one by one, shaping them, making relationships between timbral objects, and exercising great care in other dimensions of the music to insure the overall musical continuity. More contrast in the timbral dimension, less reliance upon timbre in its carrier function, need not lead to a chaos of sounds. Other musical dimensions are able to take over functions formerly given to timbre as carrier.

These two examples should remind us too that there is little likelihood of finding in any music a situation where timbre functions only as a carrier or where timbres are nothing but objects. Both functions are important in any music and there will always be a discernible residue of either function. Nevertheless, the new approach to timbres as objects represents a watershed in music. Sounds, a variety of sounds, all sounds, have become available to music. The management of this vast new vocabulary has become a central interest for musicians.

FIGURE 4

2

Some Territory Between
Timbre and Pitch

No body of theory about musical timbre exists beyond the rules of thumb and the practical advice of textbooks of orchestration, even though much of the music of modern times is more about timbre than it is about pitch. Pitch is secondary to timbre in most electronic and tape music, where various aspects of the timbral dimension are often quite precisely controlled. If composers are now articulating this timbral world it should be valuable to examine our ideas about pitch, timbre, and chord to see whether they still reflect what we are doing.

Our received theory of harmony and of fundamental pitch has its roots in the theories of Rameau and d'Alembert. Its later elaboration is built upon the psychoacoustics of Helmholtz. Doubtless much of what Helmholtz had to say is still valid (I believe it is), but a hundred years has passed since *The Sensations of Tone*, the limits of music and musical have changed somewhat, and a soundly based theory of those dimensions of music subsumed under timbre might help us to think more realistically about practical compositional problems.

We need such a conceptual framework for properly discussing the music of older composers too. Schoenberg's "Summer Morning by a Lake," from his *Five Orchestra Pieces*, Opus 16, has passages where chords are perceived as timbre, and there are a number of examples from Webern's pre–twelve-tone music. Chords approaching timbres go back to Debussy, who used them extensively. The Tristan chord is, among other things, an identifiable sound, an entity beyond its functional qualities in a tonal organization.

The decline of tonality, of root relations among chords, and of chord progression has allowed for other compositional uses of simultaneously sounding pitches. Some uses have extended traditional concepts; others have treated pitch aggregations as "composed wholes," and the term "chord" has come to be less meaningful as it is used to describe a large number of musically different situations and functions.

Nor is the sharp separation of sounds into pitches and noises, so long a fixture of musical thought, a productive dichotomy. Precision or definition of pitch can no longer be treated always as "given." We are able to compose the "pitchiness" of noise, to limit the spectral envelope to a very narrow band, or to make broad bands of noise; electronic equipment allows us to mix these at will. Some of us are hearing pitch, or pitchiness, in situations where we used to assume no particular pitch, for example, in the so-called unpitched percussion instruments.

A third gray area, between a single pitch and chord, may appear more problematical, because musicians have always been able to perceive the difference between one sound and the several necessary for a chord; yet new instrumental and electronic sounds have already turned up hints that the borderland between a single pitch (with timbre) and a chord can be musically fruitful.

I wish to look at some of these ambiguities, relationships, and transformations concerning timbre, pitch, and chord; to chart the territory; and to examine a few instances. A diagram helps to illustrate my idea (fig. 5):

FIGURE 5

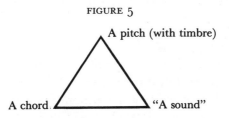

The points of the triangle represent the concepts as they are commonly used by musicians. Two of them are self-explanatory; "a sound" is meant to connote an unpitched sound without a definite pitch or pitches, for example, the sound of a bass drum.

Here is an example that brings together a number of the relationships and transformations I am interested in. J. C. Risset, who synthesized the computer-generated sounds, explains: "This run presents an attempt to prolong harmony into timbre: a chord, played with a timbre generated in a way similar to ring modulation, is echoed by a gong-type sound whose components are the fundamentals of the chord. The latter sound is perceived as a whole rather than as a chord, yet its tone quality is clearly related to the chord's harmony. The passage is as follows" (1970:A103) (see fig. 6, p. 20).

In spite of Risset's remarks I am not at all sure that I hear any of the three parts of the example as a chord. The beginning sequence might be taken as a broken chord, but only within a particular musical-stylistic framework. The

FIGURE 6

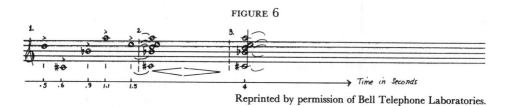

Reprinted by permission of Bell Telephone Laboratories.

second set of sounds hovers between timbre and chord—depending upon the musical context it could function as either a timbre without a single dominant pitch or a chord. The final element is strikingly different: it is definitely a pitch having a certain timbre.

Two other points could be made about the set of sounds as a whole. There is a certain unity: all the pitch elements were the same, even though one could not easily perceive individual pitches in the final part. Furthermore, all the tone qualities, whether of single pitches, of the middle part, or of the final sound, belonged to a class. There were no "foreign-sounding" characteristics in the timbral dimension—the inharmonic tone quality was not violated.

PITCH (WITH TIMBRE)→CHORD

The final part of Risset's example, a pitch having a certain timbre, is the most common type of timbre perception in music. Pitches come with timbre attached, and we are not aware of any component partials. They are fused into a single percept. But if the individual partials are loud enough they may be noticed; and by means of a harmonic tone generator, such as the instrument developed by James Beauchamp (1966), where the loudness of the individual partials may be controlled, it is easy to demonstrate transformations between pitch (with timbre) and chord. Ascending major scales are played, with different settings of harmonic levels and envelope parameters, to show the degree of fusion of spectral components as a function of the spectral and time envelopes. Depending upon the settings one hears either a fused timbre with pitch, or several individual pitches in harmonic relationship. There is no reason in principle why inharmonic generation could not be used to produce inharmonic chords.

CHORD→"A SOUND"

Transformations from chord to a fused condition where one hears "a sound" without any particular pitch have been rare in instrumental music, except in the music of Edgard Varèse. These ensemble timbres have a striking individuality, and the pitches comprising the chord are difficult to hear out. Individual instrument timbres seem to lose their identifiability—they fuse into an ensemble timbre, such as toward the end of the passage from *Hyperprism* in figure 7.

FIGURE 7

Courtesy of Colfranc Music Publishing Corporation, Copyright owners.

Varèse consistently invented vertical sound structures that were so highly individualized that ordinary terminology—chord, voicing, doubling—no longer has explanatory value. He himself used terms such as "sound-mass," "interaction," "projection" in a highly personal way, without going into technical detail in any of his published writings. It is certainly possible that each of these ensemble timbres is a composed object, that each is unique, and that there are no broad underlying principles to uncover, but since other composers have also successfully created ensemble timbres with acoustical instruments and with electronically generated sounds, it is worth our while to examine musical conditions favoring their creation.

A CLOSER LOOK AT THEORIES OF TIMBRE AND PITCH

In music theory timbre and pitch have been kept sharply separated. Pitch has been structure, timbre has been ornament. I do not believe it possible to continue the fiction that pitch and timbre are independent separable dimensions in our music, especially when any musician can cite examples of the confusion of pitch with timbre on the nuance level in musical performance, and when a number of psychoacoustic experiments support a broader, more flexible view.

In W. D. Ward's review of a number of experiments on the pitch of pure and complex tones he points out that "when comparing the pitch of a complex tone with that of a sinusoid, it is often difficult to keep from judging the more complex tone to have a slightly higher pitch, even after setting them 'equal' ";

and in discussing some experiments by W. H. Lichte and R. F. Gray he concludes that "there is a strong suggestion that some of these listeners were not keeping timbre and pitch separate" (1970:427–428).

Explanations, and, more often, unexamined assumptions about timbre, pitch, musical intervals, and simple chords, usually can be traced to the ideas of Helmholtz: that pitch equals frequency (simple tones); that when we hear the pitch of a complex tone we are hearing the frequency of the fundamental, which is often the loudest element; that we could, in principle, hear out all the partials of a complex tone; that "the pitch" is the fundamental frequency and the timbre the pattern of the partials.

Helmholtz's model presents the ear as essentially a frequency analyzer, a sort of tiny harp in the cochlea, where each string reacts to a single frequency. This model has fitted fairly well with much musical experience, and with traditional theories having to do with pitch and consonance/dissonance, especially for music of the period 1600–1900. The theory does not work so well in explaining problems of timbre; for example, it does not adequately explain the resistance to fusion of a number of simultaneous instrumental timbres, but this deficiency has often been overlooked by investigators.

The Helmholtz theory rests upon Ohm's acoustical law, which in turn is an adaptation to acoustics of Fourier's theorem. Helmholtz used Ohm's acoustical law as the fundamental idea for his own theory of tone perception, and his restatement and extension of Ohm is the usual formulation: "Every motion of the air which corresponds to a composite mass of musical tones, is, according to Ohm's law, capable of being analysed into a sum of simple pendular vibrations, and to each such single simple vibration corresponds a simple tone, sensible to the ear, and having a pitch determined by the periodic time of the corresponding motion of the air" (1877:33).

According to this a timbre is nothing more than a collection of simple tones that we usually perceive as a mass. Each of these simple tones, sine tones, is "sensible to the ear." No doubt some of them *are*, for some ears in some circumstances. Not until the experiments of Plomp (1964) was it determined that even under the best of circumstances we can hear out only the lower harmonics. The higher harmonics cannot be heard out individually. Thus, although the ear is indeed a frequency analyzer, it is an imperfect one.

This fact, in itself, need not disturb the Helmholtz view insofar as it equates timbre with spectrum and pitch with simple tone or frequency. We do hear pitch when we are presented with a simple tone, a single frequency, and we hear changes of timbre correlated with changes in the amplitude of spectral components. For music the broad outlines of Helmholtz's theory of timbre (not pitch) may be correct but incomplete. He was aware that the attack is an important cue in the recognition and identification of musical instruments, that phase, time envelope, noise bands may influence the perceived quality of musical and nonmusical sounds. Today we have different conceptions of what timbre is for, what its potential musical uses are—we think about it differently; but for Helmholtz a musical tone was a periodic tone, and he chose to limit his definition of timbre. Therefore his theory was anything but a theory of noise perception—noises had nothing to do with music. Today we know that noises are integral to musical timbre and we would like to know a great deal more about the details of noise perception. The taste of Helmholtz's time and his own preferences required for a good musical effect, "A certain moderate degree of force in the five or six lowest partial tones, and a low degree of force in the higher partial tones" (1877:362). Today we are willing to hear "good musical effect" where the timbral materials have a very high degree of force in the higher partial tones.

THEORIES OF TIMBRE AND PITCH

For Helmholtz, then, the timbre, or quality of tone, depended upon the form of vibration. Acting as a frequency analyzer, the ear,

when its attention has been properly directed to the effect of the vibrations which strike it, does not hear merely that one musical tone whose pitch is determined by the period of the vibrations in the manner already explained, but in addition to this it becomes aware of a whole series of higher musical tones, which we will call the *harmonic upper partial tones,* and sometimes simply the *upper partials* of the whole musical tone or note, in contradistinction to the *fundamental* or *prime partial tone* or simply the *prime,* as it may be called, which is the lowest and generally the loudest of all the partial tones, and by the pitch of which we judge the pitch of the whole *compound musical tone* itself. The series of these upper partial tones is precisely the same for all compound musical tones which correspond to a uniformly periodical motion of the air. [1877:22]

Thus pitch equals frequency, and the pitch of a complex tone is determined by its fundamental frequency.

We do commonly hear pitch with timbre in music—one pitch with a tone quality. Hearing out individual pitches of (lower) partials is the exception rather than the norm, and this aspect of Helmholtz's theory has continued to seem plausible (even after the little-harp-in-the-ear part of the theory has been subjected to heavy criticism) chiefly because it was consonant with so much ordinary musical experience.

Between 1938 and 1940, however, Schouten carried out a number of experiments with the aid of an optical siren that could produce any desirable periodic wave form. He first investigated the pitch of periodic pulses where the fundamental of 200 Hz was cancelled out. The pitch of this complex sound was exactly the same as when the fundamental was included! He described a periodic sound having many partials thus: "The lower harmonics can be perceived individually and have almost the same pitch as when sounded separately. The higher harmonics, however, cannot be perceived separately but are perceived collectively as one component (the residue) with a pitch determined by the periodicity of the collective wave form, which is equal to the fundamental tone" (1940:360).

Schouten's ideas were subjected to examination and criticism during the fifties, and after the usual vicissitudes of any new theoretical idea the concepts of residue and periodicity pitch are now fairly well established. In 1962 Schouten, Ritsma, and Cardozo reviewed the experimental evidence and recommended a restatement of Ohm's acoustical law:

Ohm's acoustical law, stating that the components in subjective sound analysis correspond in a one-to-one relationship with the objective Fourier components, breaks down when the spacing of these components is too narrow. It should be restated in the following manner:
1. The ear analyzes a complex sound into a number of separate percepts.
2. Some of these percepts correspond with the Fourier components present in the sound field within the inner ear. These percepts have the timbre of a pure tone and pitch of the corresponding Fourier component.
3. Moreover, there may exist one or more percepts (residues) not corresponding with any individual Fourier component. These do correspond with a group of Fourier components. Such a percept (residue) has an impure sharp timbre and, if tonal, a pitch corresponding to the periodicity of the time pattern of the unresolved spectral components. [1962:1418–1419]

A consequence of Schouten's discoveries about the relationships between groups of Fourier components was a definition of pitch which seems accurately to reflect our experience in music. "*The pitch ascribed to a complex sound is the pitch of that component to which the attention, either by virtue of its loudness, or of its contrast with former sounds is strongest drawn.* Therefore the pitch of a complex sound may be different depending upon the circumstances under which it is heard" (Schouten, 1940:361).

This excellent definition reflects our pitch/timbre experience in music, where all sounds are complex, where many pitches and many timbres are involved, and where attention is integral to the musical experience. We hear sounds and read pitch and timbre, but frequency does not always equal pitch.

The new theory of pitch based upon periodicity perception was developed further during the fifties and sixties, chiefly by the men around Schouten, but it is by no means complete, and it appears to have raised new problems while providing an explanation for others. Plomp's book on tone perception presented an excellent summary of the current state of the theory, and included a diagram that illustrates how pitch and timbre are related to the

FIGURE 8. Illustration of the way in which pitch
and timbre are related to the physical parameters of tone.

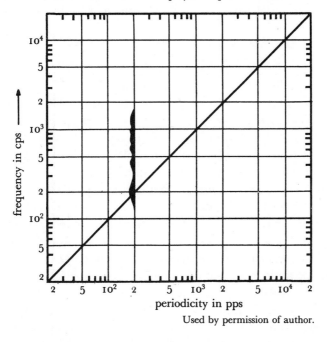

Used by permission of author.

physical parameters (fig. 8). The residue aspect of the diagram has been omitted here.

The horizontal axis represents periodicity in pps and the vertical axis frequency in cps. Since for simple tones both quantities are numerically equal, these tones are represented by a straight line with a positive slope of 45 degrees. All complex tones are localized above this line, because the period of a complex tone is equal to the reciprocal of the frequency of its fundamental. . . ."

A complex tone can be represented in this graph by plotting the repetition rate of the sound waves along the horizontal axis and the frequency spectrum measured in pass bands equal to the critical bands along the vertical axis. In fact, we need a third axis, perpendicular to the other two, to represent the sound-pressure levels in the different pass bands. This is illustrated for a complex tone of 200 cps with harmonics up to 1600 cps; the width of the line represents sound-pressure level. The psychological parameters pitch and timbre are related in a simple way to the three axes: pitch is correlated with the periodicity and timbre with the frequency spectrum. [1966:134]

The main argument against periodicity pitch is that synchronism of nerve impulses is not preserved above 3,000 or 4,000 Hz (Plomp, 1966:112–113), even though we are able to discriminate tones up to 12,000 Hz and beyond. Plomp and others have made the point that our sense of musical interval declines at about 4,000 Hz, the top of the piccolo range. Plomp even restricts his definition of pitch to the melodic aspect of tones and presents arguments that higher tones are discriminated on the basis of timbre rather than pitch.

WHY DO INDIVIDUAL INSTRUMENTAL TIMBRES RESIST FUSION?

Neither the Helmholtz theory of auditory analysis nor later writers present adequate explanations of how we are able to hear the individual timbres of musical instruments when they are playing in a group. Individual timbres are stubbornly resistant to fusion in ordinary music listening, and a composer who wishes to create ensemble timbres must take special care if he wishes to blend the instrumental timbres into a new sound.

According to Helmholtz, elements facilitating separation include temporal separation, that is, when one tone is heard for a time before being joined by another we know what we have to deduct from the compound effect; dynamic factors, such as the decay of a piano tone, or the buildup of the initial transient in brass instruments; other musical elements, such as melody and counterpoint. Of these, only the dynamic factors involving attack, decay, and modulation processes during the "steady-state" seem convincing today.

Jan Nordmark (1970) pointed out that even though all dynamic factors in a presentation of a mixture of different complex sounds are eliminated, for example, a mixture of amplitude-modulated tones, square waves, and filtered pulse trains, they still retain enough of their quality to allow identification. Whether these results hold, for, say, electronic music as presented in the concert hall, would be worth finding out. Experience with electronic music leads me to think that Nordmark has left out too many conditions—single vs. multichannel, durations, onset and offset times, mixing methods employed, pitch ranges used—for his statement to be taken as it stands.

Plomp mentions that even though the timbre of a complex tone ought to be influenced by the degree to which its harmonics are masked by other tones, we are, when listening to a concert, not aware of this effect, and he believes this should be understood as being similar to the ability to recognize familiar visual objects even when they are partly covered by other ones. He goes on to say that

hearing is a dynamic process and the various laws of gestalt psychology play an important rôle in it. All tones produced simultaneously by different sound sources are characterized by a typical time function in pitch, timbre, and loudness and this fact contributes much to their recognition. These aspects of hearing, based on central processes in the brain, are, however, beyond the scope of this study in which only some properties of isolated tones and multitone stimuli were investigated. Much more research, in particular on pattern recognition of auditory stimuli, must be done before all aspects of tone perception may become clear. [1966:136]

There is, finally, a more general idea, related to attention and to central processes in the brain, which should be mentioned here. We are interested in musical objects such as chords, "sounds," and pitches with timbre attached. If the same musical object may give rise to different perceptions, depending upon attention, musical context, and, no doubt, including our past experience and our predilictions, and if, in music, particular pitches, chords, "sounds" may sometimes be confused with each other, then somewhere there must be a decision process to resolve these confusions.

Helmholtz had nothing to say about this in *The Sensations of Tone*, but in his lecture of 1857 "On the Physiological Causes of Harmony in Music" he remarks that "We have, as it were, to distinguish between the material ear of the body and the spiritual ear of the mind. The material ear does precisely what the mathematician effects by Fourier's theorem . . ." (1857:44). We know today that the material ear is a somewhat limited frequency analyzer, but the distinction Helmholtz makes is of interest, because in his *Physiological Optics* of 1856 he discussed the relation between sensation and perception in more detail, and even though his remarks are about visual perception I believe they are pertinent to musical transformations between chord, pitch (with timbre), and "a sound."

The psychic activities that lead us to infer that there in front of us at a certain place there is a certain object of a certain character, are generally not conscious activities but unconscious ones. In their result they are equivalent to a *conclusion*, to the extent that the observed action on our senses enables us to form an idea as to the possible cause of this action; although as a matter of fact, it is invariably simply the nervous stimulations that are perceived, that is, the actions, but never the external objects themselves. But what seems to differentiate them from a conclusion, in the ordinary sense of the word, is that a conclusion is an act of conscious thought. An astronomer, for example, comes to real conclusions of this sort, when he computes the positions of the stars in space, their distances etc., from the perspective images he has had of them at various times and as they are seen from different parts of the orbit of the earth. His conclusions are based on conscious knowledge of the laws of optics. In the ordinary acts of vision this knowledge of optics is lacking. Still it may be permissible to speak of the psychic acts of ordinary perception as *unconscious conclusions*, thereby making a distinction of some sort between them and the so-called conscious conclusions. And while it is true that there has been, and probably always will be, a measure of doubt as to the similarity of the psychic activity in the two cases, there can be no doubt as to the similarity between the results of such unconscious conclusions and those of conscious conclusions. [Warren and Warren, 1968:174]

Luckily, musicians need not solve the unsolved problems of perception before music can be played or composed, although some understanding of mental and physical processes may help us to think about sounds. The views of the auditory theorists are nevertheless interesting when they throw some light, however dim, on problems of composing, performing, or thinking about music. In summarizing the ideas of auditory theory useful to musicians I would emphasize:

1. Pitch is ascribed. Attention is involved. The pitch of a complex sound (and music deals almost entirely with complex sounds) may be different, depending upon the circumstances under which it is heard (Schouten).

2. Pitch may be related to periodicity up to about 4,000 Hz, which marks the top notes of the piccolo range, and also marks the region above which our sense of musical interval declines (Plomp).

3. Timbre is related to spectrum, even though other factors, such as attack, modulation processes, phase, noise bands may also be significantly involved (Helmholtz and others).

4. The ear is a limited frequency analyzer. We can hear out the lower partials of a complex tone, but the higher partials fuse. A mass of partials which the ear is unable to resolve is heard as timbre or a residue pitch (Schouten, Plomp).

5. Complex sounds, such as the sounds of most musical instruments, resist fusion when they are heard in ensemble. No convincing single explanation has been proposed, although two writers have suggested that the ear should be thought of as a pattern analyzer, and that perhaps the solution lies in that direction (Plomp, Nordmark).

6. If central processes in the brain, attention, gestalt factors are involved, then remarks by Helmholtz about unconscious conclusions in visual perception may be extended to some aspects of music perception too, in ambiguous situations involving decisions between pitch, timbre, chord and "a sound" in music, especially today's music.

MUSICAL APPLICATIONS: PITCH (WITH TIMBRE)→CHORD

Helmholtz draws our attention over and over again to the concept that a complex tone with harmonically related partials is a chord, but he presents it intellectually, as a principle, and we understand it in theoretical terms, as a part of his argument about consonance/dissonance and tonal relationship, rather than as a practical possibility. Nevertheless, there are instances where a single pitch (with timbre) is unraveled into a perceived chord, and there are whole musics based upon these sounds.

During a four-day ceremony at the Gyutö Monastery near Dalhousie, India, in 1964, Huston Smith (Smith, Stevens, and Tomlinson, 1967) recorded lamas chanting in such a way that certain harmonics were made audible as distinct pitches accompanying the fundamental pitch. The effect is of one person singing a continuous chord. Extended chants present alternations and nuances of inflected drone chords whose fundamentals are pitched near cello low C or the B below. There is no initial doubt in the listener's mind that he is hearing a chord. Apparently the vocal cavities are adjusted in such a way that most of the time the fifth harmonic is strongly reinforced, but one may also notice higher harmonics, including a rather clear seventh harmonic, and, at times, a third harmonic. The sound is very rich, and it is often difficult to be sure whether one's primary impression is of chord or pitch (with timbre), especially in the transitive passages between relatively stable drone notes. What makes the style so fascinating is that to follow the music requires partly an overall reading of a rich, fused sound, and partly a reading of individual pitches. Sometimes a chordal quality and sometimes the "tune" that emerges from the movement between various harmonics predominates.

One could dismiss these sounds as merely remarkable instances of spiritual and physical discipline were it not for the fact that others, even Western musicians, have been able to learn elements of the basic technique without great difficulty. If it is learnable there is no reason why it cannot be used in contemporary music of the West. Two graduate students at the University of California, San Diego, Vladimir Voos and Dwight Cannon, have become quite adept at singing in this manner after several weeks of practice, and anthropologist/singer Terry Ellingson has discussed some of the problems of vocal control and the muscular adjustments involved. Ellingson reported (1970) that he has been able to produce chords using other intervals than those in the lama recordings, and that there is a considerable range available for the lower note.

Jon Glasier, also at the University of California, San Diego, has shown that a singer can select and control individual harmonics higher than the fifth or sixth, and has used the harmonics to make tunes over a drone, somewhat in the manner of a jaw harp, and he has demonstrated that two-part counterpoint is possible for one person (fig. 9).

Mr. Glasier's voice is pitched higher than that of the Tibetan lamas. Nevertheless, he is able to produce many harmonics, and the individual harmonics are easily distinguishable. We can think of his vocal apparatus as a variable band-pass filter which he uses to filter individual harmonics of different fundamentals, and the vocal apparatus as used in Tibetan technique as a set of (usually) fixed narrow band filters.

The sounds from Beauchamp's harmonic tone generator do not readily fuse unless the note length is rather short and/or the harmonics are low in amplitude. One might anticipate more fusion from a generator capable of producing more harmonics, but even with six harmonics various settings of the controls illustrate nicely the range between hearing a pitch (with timbre) and hearing a chord, and the great importance of the time scale.

FIGURE 9. Two-part music sung by one singer.

FIGURE 10. Tune with accompaniment, one singer.

The transformation from pitch (with timbre) does not appear to be limited to chords generated by the harmonic series. The partials of bells and carillons are not harmonics of some fundamental, and their characteristic timbre is related to the inharmonicity of their partials. Furthermore, it is often difficult to decide with certainty the pitch of a bell if it is heard in ensemble.

The sound of a bell depends heavily upon the weight of the clapper and the clapper material. A steel clapper will excite more strongly the upper partials, a bronze clapper will excite more lower partials. Whatever its chief pitch, one is usually strongly aware of one or more partial tones as competing pitches.

Rods also have inharmonic partials, quite bell-like, and more economical to construct. Many modern electric carillons actually consist of amplified and filtered rods.

The strings of a modern grand piano are large enough and stiff enough to act somewhat like rods. Even though their chief modes of vibration are stringlike an element of inharmonicity is introduced. While small, it is not negligible: O. H. Schuck and R. W. Young, studying piano tones (1943), were unable to find any true harmonics at all; indeed, on some tones they found that the fifteenth partial was approximately the frequency of the sixteenth harmonic of the fundamental. Harvey Fletcher (1962) found that inharmonicity was necessary if listeners were to classify his synthesized replicas as "piano-like."

Following up this idea, Frank H. Slaymaker (1970) used an IBM 360 computer to generate a variety of tones whose upper partials were stretched or compressed by some stretch exponent, producing chromatic scales analogous to the Railsback stretch, familiar from piano tuning. When the stretch exponent was 1.083333 the tone was somewhat bell-like to the ears of his listeners, and for a stretch of 1.261859 the effect was quite chime-like. The bell or chime effect was also quite marked when the partials were compressed, 0.792481, rather than stretched. In the bell-like or chime-like examples the individual lower partials can more easily be heard out; in fact, the sound tends to unravel into an inharmonic chord. Therefore, it seems that if one wished to pursue transformations from pitch (with timbre) to chord, emphasizing inharmonic relationships, these could be accomplished by acoustical means, bells, bars, plates, or through the use of either direct electronic sound generation or computer-generated inharmonic sounds.

CAN THIS TRANSFORMATION BE REVERSED?

If a pitch (with timbre) may, in certain musical circumstances, be transformed into a chord, then why not the reverse? Could one, for example, orchestrate a chord in such a way that a good listener might hear something which he would be willing to call a single pitch (with timbre)?

That we rarely hear such fusions depends upon so many conditions that I would like to defer a full discussion until later, when I examine transformation between chord and "a sound," but a glance at any textbook of traditional orchestration will show a concern with techniques which, at least peripherally, relate to chord→single pitch (with timbre).

Throughout the classic period and beyond the flute was much used to double a violin melody an octave higher. The perceived pitch was that of the violins; the flute, unheard as an independent instrument, brightened their sound.

Organ registration aims at mixing various types of pipes in much the same way. Octave doublings are commonplace. More interesting here are the doublings at the twelfth, and the effect of mixture stops, which change the quality of the compound sound.

A composer attempting to create a compound orchestral sound which might give the impression of a single lowest pitch having a timbre hopefully different than that of any of the individual instruments might try building a triad along the spacing of the harmonic series. A sketchy reading of the second part of *The Sensations of Tone* could lead him to write the compound sound in figure 11. It is not likely to work. Even with a heavy weighting of the lowest tone, individual dynamics for each group, an instruction for nonvibrato playing, and the attempt by writing a harmonic for the highest note to prevent it from too strongly capturing the attention, the desired effect is unlikely. String instruments are very rich in harmonics, and these harmonics are all added to the total sound. In addition, the sound lasts rather a long time—four seconds—allowing the listener ample time to hear out some of the pitches. He

FIGURE 11

need not hear them all, because his decision between single pitch and chord can be on the basis of "more than one pitch."

Another attempt uses flutes, because they are not as rich in harmonics as strings (fig. 12). The pitch of the complex is raised, and the time is considerably shortened. There is still no guarantee that the sound will fuse, but the sound of twenty-one flutes is not something one hears every day, and *something* about the compound sound ought to sound interesting.

Assuming that the players were able to start and stop absolutely together (unlikely), and assuming that they were able to tune, and keep in tune for the duration of the note the pure octaves, fifths, thirds of the harmonic series (impossible), a listener might still be able to separate the individual instruments through location cues provided by his two ears and the visual

FIGURE 12

evidence of a group of performers. A very reverberant environment would help to mix the compound sound. The context in which it appeared could tip the balance, and if this sound were part of a composition of mine I would explain to the conductor what I had in mind, put an accent on the lowest note and hope for the best.

MUSICAL APPLICATIONS: SINGLE PITCH (WITH TIMBRE)→"A SOUND"

Traditional musical practice made a fundamental distinction between pitches and noises, pitched and unpitched. After some experience with filtered white noise, a composer today is likely to think more in terms of a continuum ranging between sharply focused pitch, through pitch bands, to unpitched (usually broad-band) sounds. When noise bands are composed a number of new transformational possibilities arise, chiefly because many noise-band sounds can be perceived either as "sounds" or pitches (with timbre). With certain special noises there is an effect, remarkably like the double figures of visual perception, which forces us to reexamine some strongly held beliefs and assumptions about pitches and noises.

It has long been known that some whispered vowels may be perceived as single pitches. The perceived pitch closely matches the measured frequency of the second formant peak (Thomas, 1969). The first researcher in modern times to draw attention to the phenomenon was Donders, whose findings were elaborated by Helmholtz (1877:108–110). Helmholtz pointed out that certain vowel sequences could produce a series of musical tones, and he published a set of frequencies. There has always been disagreement about which vowels match which pitches, and this is to be expected, because the vocal apparatus of no two persons is identical, and because the "same" vowel may be quite different in North German, Dutch, or mid-American.

Allowing for such differences, whisper the following sequence of vowels suggested by Schouten (1962). It should produce one section of the well-known Westminster chime.

$$
\begin{array}{cccc}
\phi & I & \varepsilon & a \\
\phi & \varepsilon & I & \phi \\
I & \varepsilon & \phi & a \\
a & \varepsilon & I & \phi \\
\end{array}
$$

I had to make "ϕ" less like "deux" and more like "Dieux". For the "a" sound, as in "father", I needed to avoid "fawther." "Hit" was the word for "I," and "debt" for "ε." My set of pitches was:

$$
\begin{array}{cccc}
A\flat & C & B\flat & E\flat \\
A\flat & B\flat & C & A\flat \\
C & B\flat & A\flat & E\flat \\
E\flat & B\flat & C & A\flat \\
\end{array}
$$

I think it altogether remarkable that this sequence may be perceived *either as a series of whispered vowel sounds or as a tune!* Notice the either/or. If the reader will attempt to produce the tune from the given vowels he may find it hard to focus his attention; one listens to (pays attention to) either the pitches or the vowels. I find it difficult to avoid oscillating between modes, just as when I look at a visual double figure such as the "Peter-Paul goblet" of figure 13 I oscillate between the two ways of reading it. Other musicians have had the experience of change of mode of perception. It is certainly related to the fact that our vowel perception is categorical (Liberman, 1970), and that vowel sounds are extremely well known to us, but these observations do not explain it away.

Experience with these pitches which turn into whispered vowels should, at the very least, make us suspect that some sort of unconscious conclusion, such as that mentioned by Helmholtz, may be involved, and that, at least with certain types of sounds, we may be making decisions (they would need to be on the order of milliseconds) about whether to process the signal as pitch or vowel.

FIGURE 13. Rubin's ambiguous figure, the "Peter-Paul Goblet."

Note that a familiar tune is used. Without such a pitch structure we would perceive vowels (timbres). Thus, when we are set for pitch we can perceive it, if the second formant frequencies of our vowels are not too terribly out of tune to completely distort the melody, and when we are set for whispered speech sounds we do not hear the tune. What we are "listening for" is the vital difference. We need to know a great deal more about "listening for" if we are to gain any understanding of major concerns of our music, such as klangfarbenmelodie and ensemble timbre.

I have no explanation to offer for these auditory double figures, if that is what they are. No doubt, when explanations are forthcoming, they will prove once more that whatever is happening in our mental processing is extremely complex. My point is a musical one: certain sounds may be interpreted either as timbre (whispered vowels) or as pitch.

The narrower the bandwidth the easier it is for us to attribute a unitary pitch to a noise band, but there are differences of opinion about whether the pitch is always related to the center of the band. The results of some recent experiments using a noise band 100 Hz wide, placed between 480 and 580 Hz (a minor third from approximately B to D above A 440), show that we may also assign pitches related to the edges of the band (Carterette, Friedman, and Lovell, 1968). Changing the slope without changing the location of the center frequency (steepen one slope while gentling the other) can change the perceived pitch (Karlin, 1945). Other investigators, (Greenwood, 1968; Rainbolt and Schubert, 1968), have obtained results consistent with this. Their findings show the lower edge of the noise band to be slightly favored over the higher edge. Judging from musical experience with tone clusters over the past fifty years, these pitch possibilities are not unexpected.

My own experience with filtered noises confirms the psychoacoustic results. Ambiguity of pitch associated with noise bands was composed into my *Pacific Sirens*, for tape and instruments, and *9½ for Henry (and Wilbur and Orville)* in two ways: either to make composite, somewhat unpitched, broad-band noise sounds through mixing several discrete bands, or to make a "broad tuning," analogous to a thirty-two–violin unison, of a single pitch. I found that pitches sometimes changed in the mix-down process, that the tuning of pitch-band intervals was not an easy task, and that octaves had to be stretched larger than a two-to-one ratio to sound in tune.

There is an obvious need for more and better filters, such as a set of several variable band-pass filters with continuously variable bandwidth around a center frequency, controllable cutoff ratios, and roll-off rates. Presently one needs to patch together a variety of filters not especially well adapted to composing with noise bands. Occasionally I have used acoustical filters to make excellent pitched noises: a noise-band "fundamental" with harmonics occurs when white noise is passed through a cylindrical tube, with the amount of harmonic development controlled, as in organ pipes, by the length/width ratio.

NOISE BANDS AND TONAL MASSES

Some of the noise bands used in psychophysical experiments are computer synthesized. For example, the 100-Hz-wide sound mentioned above was made from fifty-six sinusoids randomly spaced in frequency, with average separations of about 1.73 Hz. This type of noise band has a theoretically infinite attenuation rate outside the passband. Imagine fifty-six flutists filling in the minor third from B to D and you will see that noise bands and tonal massing of the sort practiced in much recent music are two sides of the same coin.

What may be of musical significance is the average width between elements of the tonal mass. We are no longer (if we ever were) bound to think only in increments of chromatic half steps. Computer techniques allow precise control of the width between elements, easy specification of their frequencies and more sharply marked edges than is possible with any filters we have today.

When smaller increments are easily available by mechanical means, I am confident that some performers on traditional acoustical instruments (not keyboard instruments, of course) will be ready to produce any desired pitch nuance, once composers make their desires clear, and after a method of practice has been devised. "Sounds" having no definite pitch, such as those of tonal masses, are here to stay, and their musical possibilities for pitch and timbre are likely to be much more fully explored.

In the final moments of *Ensembles for Synthesiser* by Milton Babbitt we can hear a timbre/pitch transformation of particular elegance. A sustained tonal mass, at first punctuated by staccato bursts, changes gradually into a single perceived pitch. The pitch elements of the tonal mass gradually disappear until the only musical object left is the single pitch, the transition being so smooth that we are not easily aware of the precise time when we stopped reading timbre and began reading pitch. This sort of transformation could conceiveably be accomplished by an instrumental ensemble; on an instrument as programmable as the RCA Synthesizer it is relatively easy.

MUSICAL APPLICATIONS: CHORD→"A SOUND"

I started this discussion with a declaration that composers such as Schoenberg, Webern, Varèse, Debussy, and Wagner sometimes composed music where certain chords were meant to be perceived as timbre. It is time for a closer look at some of their methods.

The third movement of Schoenberg's *Five Orchestra Pieces*, Opus 16 (1909), is called "Summer Morning by a Lake." It has a subtitle that could have been the original name of the composition, "Colors." Harmonic and melodic motion is much curtailed, in order to focus attention on timbral and textural elements (fig. 14). Schoenberg's two notes to the conductor reveal something of his conception of the sound he desired:

It is not the conductor's task . . . to bring into prominence certain parts that seem to him of thematic importance, nor to subdue any apparent inequalities in the combinations of sound. Wherever one part is to be more prominent than the others, it is so orchestrated, and the tone is not to be reduced. On the other hand (the conductor) must see that each instrument is played with exactly the intensity prescribed—in its own proportion, and not in subordination to the sound as a whole.

The change of chords in this piece has to be executed with the greatest subtlety, avoiding accentuation of entering instruments, so that only the difference in color becomes noticeable.

The 1949 revision of the score dispenses with the first of the two notes and adds: "The conductor need not try to polish sounds which seem unbalanced, but watch that every instrumentalist plays accurately the prescribed dynamic, according to the nature of his instrument. There are no motives in this piece which have to be brought to the fore."

The method of the piece is to change the instrumentation of a chord, or a single pitch, in a structured way. The changes are usually linked, that is, the second instrument or instrumental group begins to play while the first is still playing. Figure 15 shows the main color changes of a few measures. Notice that the two groups (1: flute, flute, clarinet, bassoon, and 2: English horn, trumpet, bassoon, French horn) each include a bassoon. The viola and contrabass patterns make a complementary subsystem of timbre alternation which helps to link the alternations of the two main groups. In measure 229 new, well-contrasted pitch and timbre elements enter. Even though the play of timbre change becomes more complex, the various groups are kept as distinct timbral entities, no mean compositional achievement.

FIGURE 14

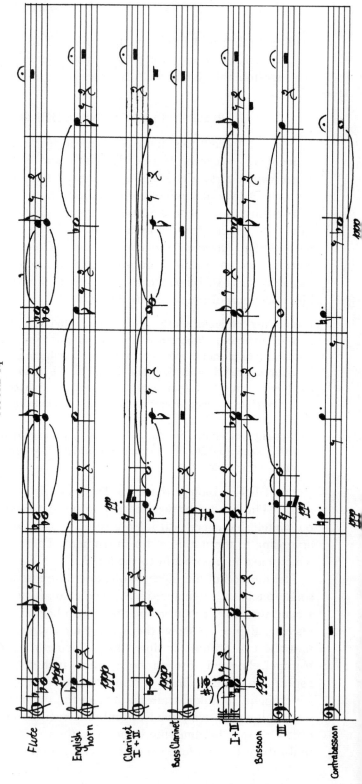

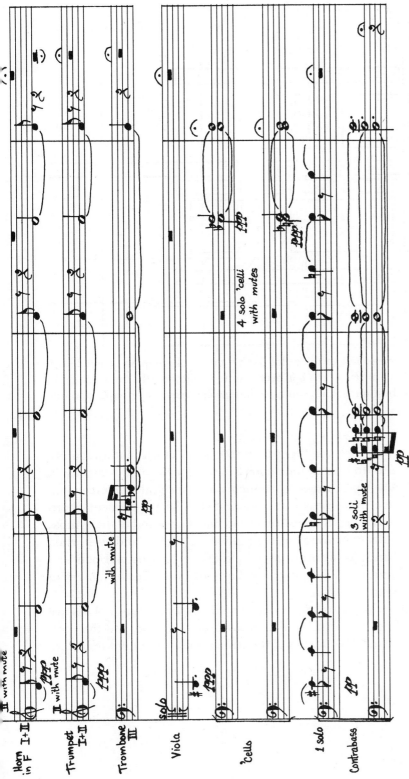

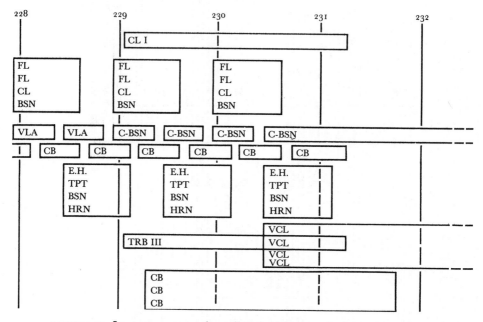

FIGURE 15. In measure 230 the cellos double the wind group an octave lower. The part for Bsn. III in measure 229 has been omitted, following the later edition of the score. Note the beginning and ending relationships between groups, and the lengths of the pauses.

Webern's *Six Pieces for Orchestra*, Opus 6 (1909), is full of fascinating sounds. Many of the five- , six- and seven-tone chords fuse enough to be heard as "a sound," especially when percussion instruments add spectral reinforcement. Figure 16 is a passage from the second movement. The ten-element chord cannot easily be analyzed into its constituent pitches. The sound of the cymbal roll adds to the difficulty of hearing out the individual pitches, and even when the cymbal roll is absent, as in measures 25 and 26, the trilling clarinets and trumpets, and the pizzicato violins, composed as part of the percussion group against the winds, promote fusion.

A reduction of the wind parts brings out some interesting details: oboes are enclosed by the trilling clarinets; horn is enclosed by trumpets, and the horn note is trilled by trumpet. The pitches of the strings double most of the pitches of the winds (fig. 17).

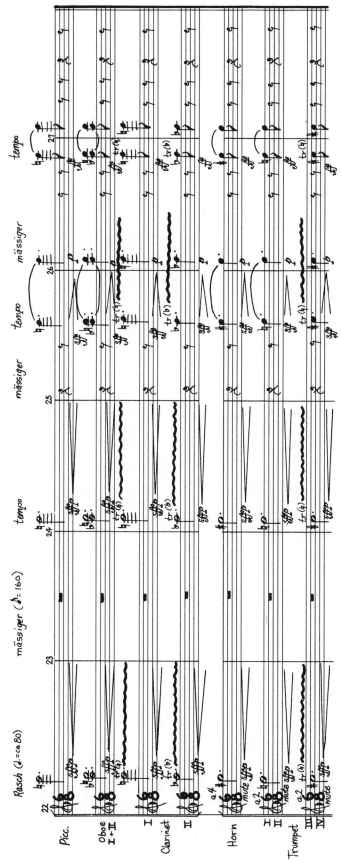

FIGURE 16

FIGURE 16 (*continued*)

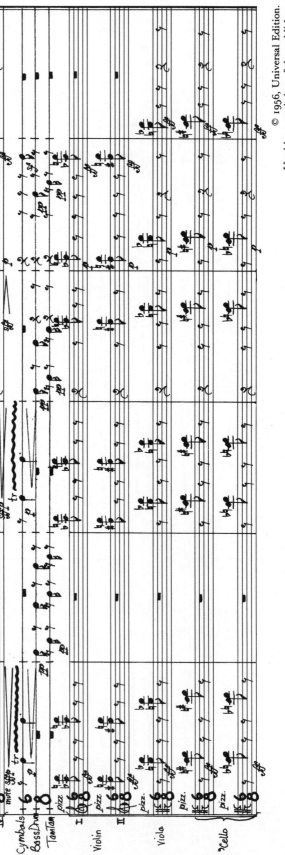

FIGURE 17

Stravinsky's treatment of chord and "sound" is somewhat different from that of Schoenberg and Webern in that (at least in his pre–twelve-tone compositions) he often attempted to preserve certain traditional aspects of chord function, and at the same time to fully characterize the vertical aggregate as sound. Walter Piston has a telling story:

In the course of a talk about Oedipus Rex, at one of those meetings, an observation that he [Strawinsky] made threw a bright light on a most important aspect of his artistic ideals. He said, "How happy I was when I discovered that chord!" Some of us were puzzled, because the chord, known in common harmonic terms as a D-Major triad, appeared neither new nor complex. But it became evident that Strawinsky regarded every chord as an individual sonority, having many attributes above and beyond the tones selected from a scale or altered this way and that. The particular marvelous combination of tones in question owed its unique character to the exact distribution of the tones in relation to the spaces between them, to the exact placing of the instrumental voices in reference to the dynamic level indicated, and to the precise moment of sounding of the chord. [1947:256]

The chord which opens the "Danse Sacrale," from *Rite of Spring* (fig. 18), is characteristically "a sound" and more than "a sound": it is scored for fusion; its brevity allows little time for hearing out individual elements; and it is reiterated; yet the perfect fifth from the bass (cellos) insures a strong intervallic/harmonic component.

FIGURE 18

A later work, *Agon* (1954–1957), has a rich variety of sounds bearing the special Stravinsky stamp, such as measure 7, from the opening Prelude, with its unusual voicing of horns and double reeds (fig. 19). Precise balance of the individual instrumental sounds, carefully calculated dynamics and accents transform the very familiar sounds of oboe, horn, and English horn into something new and memorable.

FIGURE 19

Stravinsky learned a great deal about sounds and their disposition from the music of Debussy. Listen, for example, to the opening of the second part of *Rite of Spring*. The debt is unmistakable. To say that his attitudes toward voicing and instrument registration owe much to the innovations of Debussy detracts not at all from Stravinsky's own achievement. What he carried away he made his own.

Debussy usually carefully raked over the markings of his craftsmanship, but his methods were subtle and detailed. In figure 20 notice how refined the swirl of sound in the measures from *Jeux*, a work exactly contemporary with *Rite of*

FIGURE 20

Spring. Everything happens very quickly: the first dark sound of the contrabasses and bassoons, followed, on the second eighth, by the sharp attack of the cellos and violas masking the quiet entry of the horns, followed by the very bright chord of muted trumpets, violins, flute/piccolos, and cymbal roll. Notice that the dynamics reflect the separate groups, and that (what a magical idea!) the horns swell and slightly overlap the entry of the final group, while the bassoons crescendo on the final eighth note of the measure. Notice too, how the cymbal continues throughout the falling motion of the piccolos, flute, and violins.

Flickering timbre changes, such as one hears in this passage from *Jeux*, and the mixing in of noise elements for nuances of timbre coloration are now a part of the technique of composers who may not be aware of what they owe to the discoverer. In particular, the idea of fast-changing, individualized, sharply contrasted dynamics within a short time span was to be heavily exploited later, first by Varèse, then by most of the composers of the fifties and sixties.

We may infer, from these excerpts from Schoenberg, Webern, Stravinsky, and Debussy, that whether we are likely to hear a compound ensemble sound as a single fused timbral entity or as discrete pitches of a chord depends overwhelmingly upon the musical situation:

1. If simultaneous sounds are too short they cannot easily be perceived as chords and are likely to be read either as "a sound" or even as a single pitch (with timbre). The measurement, "too short," depends upon the time context: tempo, articulations.

2. Familiar chords make for easy decisions. Unfamiliar chords or aggregates of pitches are potentially more ambiguous—and more interesting.

3. If some individual pitches can be heard out we are likely to perceive chord, even though the sound may contain many more.

4. If the component pitches are unsteady, or if one instrument is too loud in relation to the others, the compound sound will not fuse. In fact, the various instrumental timbres resist fusion, and special steps must be taken to voice and balance the ensemble sound.

5. When individual instruments begin and end together fusion is more likely.

6. Seating arrangements, hall characteristics, microphone placement contribute heavily to fusion or separation, as does the position of the listener.

With so many variables one need not expect all vertical aggregates, even when well blended, to have the quality of fused ensemble timbres. Obviously there are many degrees of blend between separation and fusion. Therefore, I would like to limit the term "fused ensemble timbre" to those sounds that show:

(*a*) a (precarious) balance of forces, where

(*b*) individual instrumental sounds lose their identifiability, and where

(*c*) an unexpected, or striking or otherwise memorable fused sound is in the perceptual foreground.

If context, position, and time are so important, then the making of such fused ensemble timbres is a compositional task—they do not happen automatically, they need to be composed. One composer, above all others, has used them extensively and made them integral to his musical conceptions—Edgard Varèse. They appear in every composition for ensemble and orchestra from *Hyperprism* (1923) to *Déserts* (1954).

FUSED ENSEMBLE TIMBRES AND THE
SOUND-MASSES OF EDGARD VARÈSE

Varèse's ideas about the uses of timbre are already well formed in his earliest published works, *Hyperprism* (1923) and *Octandre* (1924). The fused ensemble timbre three measures after rehearsal number 2 in *Octandre* (fig. 21)

FIGURE 21

Courtesy of Colfranc Music Publishing Corporation, Copyright owners.

is comprised of horn, trumpet, and trombone. Notice that the dynamics of this fused sound are composed together, against the dynamics of the other instruments.

Between rehearsal numbers 5 and 6 of the same composition there are seventeen measures of rhythmic oscillation between two seven-element ensemble timbres, contrasted by accent, register, and dynamics. These strong contrasts are reinforced by allowing the appearance of the "melodic" E♭ clarinet only with the softer of the two ensemble timbres. Repetition enhances the effect of fusion (fig. 22). But the most interesting aspect of this interplay comes from pitch interchange (fig. 23). Trumpet and flute share the high F, oboe and trombone the G, flute and trumpet the F♯, bassoon and horn the C♯, making a delicious confusion for the ear. The radical register shifts increase the number of available timbres, especially for flute, oboe, and horn.

Also from *Octandre* comes that remarkable sound from the last movement, the sixteen measures of ensemble timbre under the trumpet solo, beginning at rehearsal number 2. At rehearsal number 3 the steep crescendos in each measure introduce another gradient of timbre nuance, and demonstrate once again that timbre, including its ensemble forms, is *not* invariant under changes

FIGURE 22

Courtesy of Colfranc Music Publishing Corporation, Copyright owners.

FIGURE 23

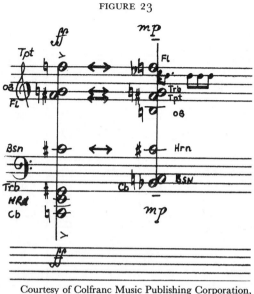

of loudness (fig. 24). The ensemble timbre changes its quality with the crescendo. The gap between the lower and higher instruments is filled by the strong second harmonics from contrabass, bassoon, and trombone. (fig. 25). The flute can add little to the crescendo, but the E♭ clarinet is almost as powerful here as the trumpet, and the trumpet itself, if played at a true *FFF*, creates so many intense high partials that they "splatter" into the total sound and help the fusion of the other instruments.

Varèse conceived his music, and talked about it, in spatial terms. Sounds were physical for him in the sense that they took up space, and had shapes and positions in space. They were concrete, touchable objects. In a conversation with Gunther Schuller sometime during the sixties he said,

I was not influenced by composers as much as by natural objects and physical phenomena. As a child, I was tremendously impressed by the qualities and character of the granite I found in Burgundy, where I often visited my grandfather. There were two kinds of granite there, one grey, the other streaked with pink and yellow. . . . And I used to watch the old stone cutters, marveling at the precision with which they worked. They didn't use cement, and every stone had to fit and balance with every other. So *I was always in touch with things of stone and with this kind of pure structural architecture*—without frills or unnecessary decoration. All of this became an integral part of my thinking at a very early stage. [Schuller, 1965:34; italics mine]

In the light of these remarks, and keeping in mind that Varèse's favorite image for musical organization was crystal structure, one gets a glimpse into

FIGURE 24

Courtesy of Colfranc Music Publishing Corporation, Copyright owners.

FIGURE 25

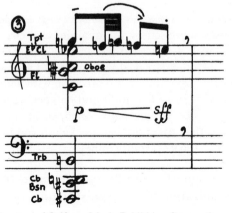

Courtesy of Colfranc Music Publishing Corporation,
Copyright owners.

FIGURE 26

his fundamental attitudes toward sound from something he said in a lecture given at Princeton University in 1959. "There is an idea, the basis of an internal structure, expanded and split into different shapes or groups of sound constantly changing in shape, direction and speed, attracted and repulsed by various forces" (Chou Wen-chung, 1966:16).

These remarks make excellent sense as soon as one stops trying to interpret them within traditional music/theoretic categories. One has to forget about consonance/dissonance, melody, harmony, and understand that, when he says "different shapes or groups of sound," he is speaking seriously and precisely, and that he is grappling with musical dimensions for which no theory existed. By applying spatial concepts in a very radical way he at least provided himself with a framework for thinking about what he was doing, and he may have provided us with notions and ideas more pertinent to the discussion of timbral organization than we have thought.

One can show how he used "shapes or groups of sound" structurally. His central term was "sound-mass," and I believe a fused ensemble timbre is a variety of "sound-mass." There are two distinguishable sound-masses starting at measure 5 of *Intégrales* (fig. 26). The first, comprising piccolos and B♭ clarinet, is a fused ensemble timbre; the other, three trombones, may hover closer to chord. The sound from measure 4, produced by gong and tam-tam, is another sound-mass, and a fused ensemble timbre. The two wind instrument sound-masses tend to merge during measure 6. Their fusion is promoted by the broad band foreground noise of the tenor drum. Thus, "different shapes or groups of sound constantly changing in shape."

FIGURE 27

The B♭ of the E♭ clarinet is rhythmically, accentually, and dynamically differentiated from the other musical members, but it has strong timbral affinities to the piccolo/clarinet group: in that particular range its quality is such that it fuses easily with the piccolos and the other clarinet; in fact, the combination of E♭ clarinet with piccolos or flutes is almost a cliche of Varèse's vocabulary of ensemble timbres. Thus, "attracted and repulsed by various forces."

Varèse is careful to establish layers of sound, and these layers occupy positions in pitch space. There are bands of sound, strata. The progress of the music depends heavily upon the flowing together (fusion) or separation of the strata. When each stratum is itself a fused ensemble timbre then the musical structure is one of merging, separating sound—organized sound. "The role of color or timbre would be completely changed from being incidental, anecdotal, sensual or picturesque; it would become an agent of delineation like the different colors on a map separating different areas, and an integral part of form" (ibid.:12).

A more complex use of ensemble timbre appears at measure 25 of *Intégrales* (fig. 27). The staggered entries of trumpets, E♭ clarinet, oboe, and horn, so much a part of Varèse's idiom, do nothing to enhance fusion; they counter and disrupt it. This "shingling" of the individual sound elements is melodic in the sense that pitches are placed sequentially; but shingling also promotes the easily heard disposition and registration of individual elements in pitch space, and, more importantly, it is a way of presenting unfused single timbres. For Varèse is working not only with fused ensemble timbres, but with the whole range between separation and fusion, and movement between these states is fundamental to his art.

FIGURE 28

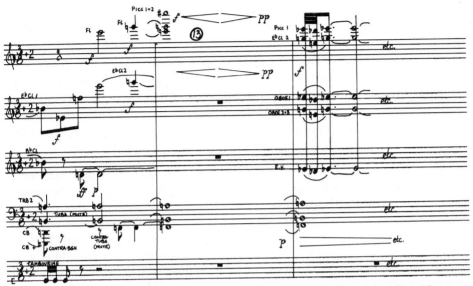

Courtesy of Colfranc Music Publishing Corporation, Copyright owners.

When new instruments will allow me to write music as I conceive it, taking the place of linear counterpoint, the movement of sound-masses, of shifting planes, will be clearly perceived. When these sound-masses collide the phenomena of penetration or repulsion will seem to occur. Certain transmutations taking place on certain planes will seem to be projected onto other planes, moving at different speeds and at different angles. There will no longer be the old conception of melody or interplay of melodies. The entire work will be a melodic totality. The entire work will flow as a river flows. [Ibid.:11]

Measures 24 through 29 from *Intégrales* (fig. 27) present several separate fused ensemble timbres which finally cohere into a single sound. Notice the pitch range: Varèse has discovered the contrabass trombone, and is using it to build low-pitched strata. By measure 27 the two higher sound-masses are overlapping (perhaps Varèse's penetration), and after the *PP subito* of measure 28 the whole ensemble fuses. The E♭ clarinet functions in a manner similar to that of measures 5–7. "In the moving masses you would be conscious of their transmutations when they pass over different layers, when they penetrate certain opacities, or are dilated in certain rarefactions" (ibid.:12).

We may never know precisely what Varèse meant by "penetration," "opacities," or "rarefactions"—especially "rarefactions"—but measures 24–29 give us an opportunity for at least an intuitive grasp of their referential force.

Like *Intégrales, Arcana* (1927), for large orchestra, makes much use of fused ensemble timbres in the high register in opposition to low, widely spaced brass. The elegant passage just before rehearsal number 13 (fig. 28) uses the power of two E♭ clarinets to kick off the sound, leaving hanging the ensemble timbre of two flutes and piccolos. As is his practice, Varèse reinforces the contrast between the flutes/piccolos and the low brass sound by giving separate dynamics to each group.

His need to extend the available pitch space, and the number of musical situations where stable, high, loud pitches were useful to him, led Varèse to explore the musical potential of such electronical musical instruments as were available, or talked about, during the twenties and thirties: oscillators, the theremin, Ondes Martenot, and others. Busoni (1911) had called attention to the instrument developed by Cahill, seeing it as a way of escaping the confines of the system of tempered tuning. Already in the thirties Varèse was dreaming that

the new musical apparatus I envisage, able to emit sounds of any number of frequencies, will extend the limits of the lowest and highest registers, hence new organizations of the vertical resultants: chords, their arrangements, their spacings, that is, their oxygenation. Not only will the harmonic possibilities of the overtones be revealed in all their splendor but the use of certain interferences created by the partials will represent an appreciable contribution. The never before thought of use of the inferior resultants and of the differential and additional sounds may also be expected. An entirely new magic of sound! [Chou Wen-chung, 1966:12]

Varèse had the inventor design a pair of Ondes Martenots for use in his composition *Ecuatorial* (1934). This instrument is capable of producing continuous glissandi and stable high pitches. Since the Ondes Martenot is not an easily procurable instrument (I have never seen nor heard one live) two oscillators are often substituted. In measures 15–19 from *Ecuatorial* (fig. 29), Varèse uses them much as he used piccolos and flutes in his earlier compositions. Unpitched percussion makes a wide-band noise screen, and the separate dynamics for one of the oscillators separates it for a moment, in measure 18, from the total sound.

Ecuatorial is a watershed work, the first of Varèse's compositions to mix electronic and acoustical instruments, the first to use the piano as a special sound source with attack characteristics that could be imposed on other instrumental sounds, the first to use an extended repertory of vocal sounds organized along timbral lines. Many of the new ideas scattered throughout *Ecuatorial* sprout luxuriously in *Déserts* (1954) and *Nocturnal* (1961, incomplete).

Déserts, which carries his ideas about the organization of sound-masses far ahead of any previous work, is Varèse's masterpiece. Its timbral organizations are extremely varied and refined. Varèse's particular version of klangfarbenmelodie is more richly developed than in any other work, and one finds everywhere a remarkable control over nuances of attack and decay. Fusion and separation are much more a continuous process, and the dynamics are a completely integral formal element, often carefully notated by individual

FIGURE 29

Courtesy of Colfranc Music Publishing Corporation, Copyright owners.

beats. Study of extended passages, where fusion and separation are composed on a timbral continuum, would take us too far from my purpose here, which is only to describe fused ensemble timbres and to provide a modest display of examples from the literature. So one short passage from *Déserts* will have to suffice—of particular value because it presents two pairs of fused ensemble timbres. Measures 304–307 and 308–309 (fig. 30) have the same pitch material and approximately the same instrumental registrations. The differences in the pair, except for the flute and piccolo high notes in the first, are very much a matter of the detail of the shingling: the approach to the final sound is different in each. One should carefully study which instruments enter individually and which enter in pairs or larger groups. Those that enter together will tend to fuse at the outset; more, at least, than instruments which enter individually. Individual enterers will retain their separated status for a longer time.

Measures 310–312 and 313–314 (fig. 31) are another pair of fused ensemble timbres. Differences between them are quite slight. The horn note, well separated by pitch and dynamics, is handled slightly differently in the second instance, and the addition of the low cymbal sound may change the total impression ever so slightly, but the chief difference revolves around whether the second note to enter is C or C♯, and whether the C and the high D enter together.

FIGURE 30

Courtesy of Colfranc Music Publishing Corporation, Copyright owners.

Stravinsky has called attention to the importance of the interludes of organized sound:

As an electronically organized and electronically produced sound composition, *Déserts* was possibly the first piece to explore the liaison characteristics between live instrumental music and electronically recorded sound. I refer to the transitions between the four instrumental portions and the three taped segments of electronically organized sound which connect and interpolate them; these transitions, exploiting the border country between the live and the electronically attenuated suggestion of the live, are, I believe, the most valuable development in Varèse's later music. [Stravinsky and Craft, 1963:61]

Pitched materials are more often than not chosen from a continuum. As Varèse said, "Flowers and vegetables existed before botany, and now that we have entered the realm of pure sound itself we must stop thinking in the frame of twelve tones" (ibid.:59).

Sound-masses are handled directly in the tape medium, and there is a great variety in sound-mass relationship. Movement in space is created by channel techniques and closeness/distance by control of reverberation and by the type of sound employed and its attack character.

The taped segments are really a kind of mirror world of intimations, echoes, and transformations, each subtly different from the others. The third interlude includes the widest range between pitched and unpitched sounds, and exploits the potential of the "other orchestra" as a ghostly double. The sounds include

FIGURE 31

Courtesy of Colfranc Music Publishing Corporation, Copyright owners.

extensions, evocations, and developments from the world of acoustical instrument sounds. Six places where the stage orchestra and organized sounds meet are critical points for the structure. At those moments Varèse is more interested in relatedness than in contrast; he composes them as transitions and manages them in such a way that when the stage orchestra returns one feels a (delayed) shock of recognition as though returning home after a journey to some distant place.

With the composition of *Déserts* Varèse achieved many of his lifelong desires: to compose sound-masses projected in space, free from the straitjacket of tempered tuning, where no one parameter should enjoy privileged status, and where the entire work is an intermixed unity which "flows as a river flows." His vision about composing the sound itself, rather than certain abstractions more or less related to sound, opened up the sound world we have been colonizing ever since.

3
Timbre and Time

The most musically significant thing about sounds, timbral objects, is not that they are recognizable, identifiable, nor that they are multidimensional wholes, individual and various: it is that they exist in time, have a shape in time, exhibit changes during their time course, and still retain their identity.

If composers have shown interest in composing the whole sound, and have tended to turn away from composing the kind of instrumental music where musical timbre is conceived in terms of invariant qualities accompanying pitch in favor of conceptions where sounds more often function as timbral objects, then anything that can be learned about the nature of these objects ought to be of some practical value; and it ought to be worthwhile to examine the time course of single sounds, the microstructure of their beginnings, middles, and endings to determine how much of it can be composed for traditional instruments and/or other media.

Musicians tend to think of known sounds as fixed or stationary from beginning to end, or at most waxing or waning in loudness, whether timbre is functioning as carrier or object. We have not paid much attention to timbral nuance during the time course of a sound in Western music, partly because we have had no practical notation. Nuance on the microlevel has been left to the performer by default.

Yet every sound is in motion. It is shaped in time, and its shape is perceived in time. Motion pictures of sounds are remarkably descriptive, and the best visual diagram would be cinematic, but a second best visualization will have to do until such time as we have moving books. The graph in figure 32 is a time plot of the first six partials of a middle C played on a Baldwin grand piano. The six partials of the sound do not decay at the same rate. Indeed there are spectacular changes of level for some of the partials, especially during the first third of the tone. The points of analysis are spaced at intervals

FIGURE 32. Relative level vs time for the first six partials of the tone C (middle C).

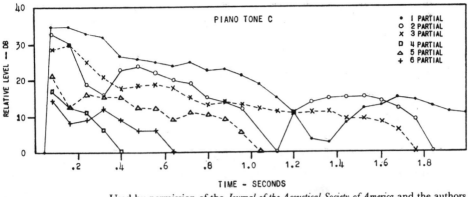

Used by permission of the *Journal of the Acoustical Society of America* and the authors.

of 80 milliseconds. If readings had been taken at closer time intervals we could expect even more level changes among the partials.

These continuous changes of intensity among the partials may be expected to have an effect upon some aspect of the timbre. At the same time, amidst all the change on the microlevel there is a group tendency—loud beginning and gradual decay. It is not surprising that we are able to perceive the overall tendency—the time envelope of the sound. The surprise is that we are so sensitive to such details as the decay rates of the separate partials and the attack time of the sound, and yet unconscious of our sensitivity.

Fletcher and his associates (1962) tested changes in attack and level relationships among harmonics using musicians and nonmusicians, where the listeners were asked to judge which tones were most piano-like. For the G two octaves below middle C they found that attack times longer than 90 milliseconds were questionable (i.e., not piano-like). In order for that tone G to be considered piano-like by the jury the difference in level of the partials had to be on the order of 2.5 to 1.5 db as the partial numbers ascended, although the amplitudes of the first five or six partials could be shifted around to some extent. Different tones required different roll-off rates.

Very slight departures from real piano tones were discriminated by Fletcher's listeners. They judged "naturalness," presumably comparing the test sound to remembered piano sounds. As far as music is concerned the tones that were not piano-like might turn out to be just as useful as the accurate reconstructions—that is beside the point. What is important is that Fletcher's listeners—and by extension all of us—made an overall judgment that took into account a variety of happenings on the microlevel. As Schouten said:

Evidently our auditory system does carry out an extremely subtle and multi-varied analysis of these elements, but our perception is cued to the resulting overall pattern.

Acute observers may bring some of these elements to conscious perception, like intonation patterns, onsets, harshness, etc., even so, minute differences may remain unobservable in terms of their auditory quality and yet be highly distinctive in terms of recognizing one out of a multitude of potential sound sources. [1968:90]

These small differences in the time course of a sound are extremely important for music. As long as composers dealt only with acoustical instruments and human performers there was no difficulty in introducing small differences—the nuance level has always been an important domain of performers. In electronic, tape, and computer music the microlevel either must be composed or decisions must be made about the level of detail worth bothering with. The temporal "grain" of most of the electronically generated and computer-produced music (which willy-nilly reflects the design philosophy of the engineering systems) is far too gross, and an important purpose of this book is to suggest to engineers and electronics specialists some design requirements for more sophisticated instruments.

ATTACK AND TIMBRE

Almost fifty years ago Stumpf discovered that the recorded sounds of musical instruments lose their identifiability when the attack is artificially suppressed. His experiments have been repeated and extended many times since, and it is now widely accepted that the beginning of the sound, the initial transient, is essential to its timbre. For example, if one records a four- or five-second piano tone on tape, then listens to the reversed sound, it will not sound piano-like. Or if one records the sound of a bell by turning up the record gain immediately after the clapper has struck its quality will be strikingly altered.

As far as I know no explanation has been advanced for this phenomenon, except the general one that the human auditory apparatus is especially responsive to change. It is known that under certain circumstances we can distinguish differences between sounds that are only 1.5 milliseconds apart (Patterson and Green, 1970), and it is also known that the attacks of musical instruments are from a few milliseconds to as much as 500 milliseconds long, depending upon definitions of terms and techniques of measurement. A sound has to have a spectral envelope of some sort if we are to hear timbre at all, but the time envelope of the sound, especially the very beginning of its cinematic development, is equally important.

Saldanha and Corso made an extensive study "to evaluate the relative importance of harmonic structure, frequency, vibrato, transient motion, both initial and final, and steady-state duration as timbre cues in the absolute judgement of musical tones" (1964). Ten instruments were used: violin, cello, flute, oboe, bassoon, clarinet, alto saxophone, French horn, trumpet, and trombone. Each experienced player recorded single tones nine seconds long (the authors say nothing about type of articulation or dynamic level), with and without vibrato, at three pitches, middle C, the F above, and A 440. (This choice of pitches presents certain problems for cello, bassoon, and trombone— perhaps also for French horn. If these instruments had been playing in more typical ranges listeners would probably have made higher identification scores.) The raw tapes were cut and spliced in such a way that listeners could be presented with either the complete tone, a three-second section from the beginning, or the middle, or the end, or finally a three-second composite made from the beginning, a shortened middle, and the end. (The authors do not give details about the splicing—butt, 45°, or other—and they do not tell us whether the composite tone had any perceptible discontinuities.)

Their results show that the tape segment consisting of the initial transients and steady state lasting three seconds was identified correctly more often (47 percent) than any other condition, including the nine-second complete tone (41 percent). The least number of correct identifications was for the last part of the tone with the final transients (32 percent).

WHAT HAPPENS DURING THE ATTACK?

Any tone having a finite length must, according to Fourier analysis, contain more than a single "frequency." This is because it must begin and end. No matter how gradually a tone is turned on its onset is accompanied by other frequencies. These sidebands are modulation products of the change of physical state from off to on and are called transients. Their distribution around any single oscillating frequency can be mathematically ascertained through the Fourier integral.

With time intervals on the millisecond level (which include many musical attacks) these extra frequencies become important. In the case of tones a few milliseconds long the transient frequency spreading is broad enough so that we perceive only a click, even though the input stimulus is a sine tone of a single frequency.

The more sudden the onset the more the frequency spread. Figure 33 shows the time course of a sudden onset, say a tuning bar struck with a hard stick. The onset of the tone shows the frequency spreading. If broad enough we will hear this as a click. Now if clicks are perceived with extremely short sine tones then we can also expect clicks from complex tones having many frequency components. Listen closely to percussive and pizzicato sounds and you will sometimes be able to hear such a click. This frequency spreading is a consequence of the inertia of any physical system, and we can expect it to be noticeable to some degree in many, perhaps most, musical attacks.

FIGURE 33

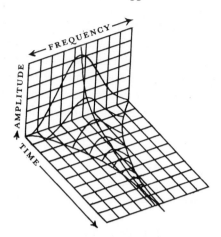

But a musical attack is much more than a simple consequence of physical inertia, and not all instruments are struck or plucked. Those excited by breath or bow have much longer attack times and therefore less of the click-causing frequency spreading. Families of instruments, for example, buzzed-lip instruments, have more or less characteristic attacks; a member of an instrument family, for example, a trumpet, will have further individual attack characteristics; and a particular performer will show additional characteristics.

The pattern of onset of the individual partials in members of the brass family is different from that of other instrument families, and patterning is extremely rich. Consider the graph in figure 34, the first 100 milliseconds of a trumpet playing an F above middle C, analyzed by Backhaus (1932). The beginning of the tone is less abrupt than that of the piano tone, although there is a first peak well before 20 milliseconds have elapsed. The harmonics rise individually and some of the lower harmonics rise fractionally earlier. There are two parts to the attack, a peak at the beginning, then a dip and swell, a pattern quite characteristic of many brass tones.

Backhaus gives no information about the type of attack, nor about whether the sound was played softly or at some higher dynamic level. We know from musical experience, however, that performers have considerable control over musical sounds through the manner of attack.

Luce and Clark made a study of the durations of attack transients in which they defined the duration of the attack as that period of time from the beginning of the signal to the point at which the magnitude of the amplitude envelope is 3 db below that of the steady state. They had each performer play scales throughout a large part of his instrument at dynamic markings of *pianissimo, mezzo forte,* and *fortissimo* (as interpreted by the performer) and at various speeds, but they gave no instruction about manner of attack. They found that the length of the attack depends upon the particular instrument,

FIGURE 34

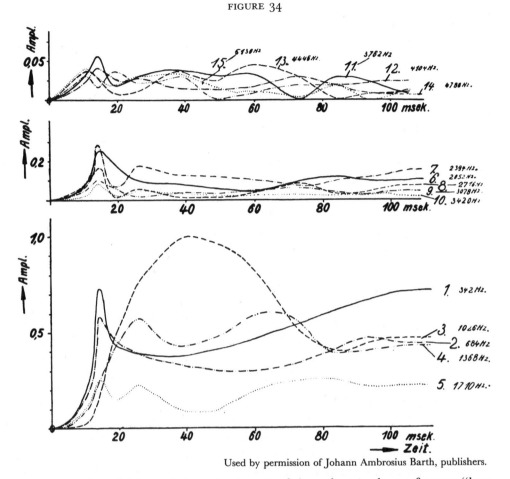

Used by permission of Johann Ambrosius Barth, publishers.

upon the pitch (higher pitches, shorter attacks), and upon the performer, "but very little on the dynamic marking at which the instrument is sounded or the duration of the notes played . . ." (1965). This finding does not conform very well to the experience of musicians: that even within a given dynamic level the length of the attack may be varied to some degree, and that sharp attacks are shorter than the gentler varieties.

Using the same method of measurement for the duration of the attack, Melka (1970) found that the manner of starting the tone changes the length of the attack. He had professional orchestra musicians play the note C at available octaves in two ways: a sixteenth note played with a sharp attack (*hart*) and a quarter note played with a gentler attack (*weich*). He gives no tempo markings for the notes but says that the sixteenth is a "very short" note and the quarter a "moderately long" note. His results for brass instruments are given in figure 35 and confirm what musicians have been practicing for several hundred years: players are able to control attack length to a

FIGURE 35

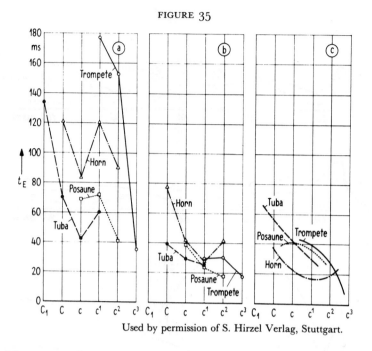

Used by permission of S. Hirzel Verlag, Stuttgart.

considerable extent. Column *a* shows gentle attacks, column *b* sharp attacks, and column *c* Luce and Clark's results.

The attacks of musical instruments are far more individualized than the *weich* and *hart* of Melka's experiment. Instruction books and instrumental methods devote more space to bowings, tonguings, fingerings, and touches than anything else. The student is taught to produce a variety of attacks on demand, throughout the compass of his instrument and at all possible dynamic levels. In recent years the vocabulary of attacks has increased. Many professional players are able to produce a far greater variety of attacks than composers have been able to symbolize. Unless such performers are given specific directions they will improvise attacks to conform with their own conceptions of what the context of the music appears to require. For many musical situations this may be desirable, but composers who might wish to notate attack qualities more specifically have had at their disposal only accents, dynamic indications, and/or discursive verbal instructions.

Donald Martino has suggested how modified phonetic symbols could be used to designate "more or less equivalent effects that can be produced by various musical instruments" (1966:55). In addition to the usual /t/ and /k/ for tongued attacks he lists /b/, /d/, /g/, /h/, /j/(as in yes), /l/, /m/, /n/, /p/, and /w/. He also suggests several vowels. This is a large step toward an accurate notation of timbre, firmly based upon human articulatory possibilities and a simple, easily understandable symbol system. Phonetic notation of the attack is immensely practical for winds; we may yet find a way to adapt it for strings.

Martino wished to create a symbol system that would serve for both attacks and decays and could replace the present morass of dots, dashes, and lines we have inherited. The difficulty with his suggested symbology (see fig. 36a) is that, except for the /h/ symbol, players are likely to decode the dots, dashes, and carets in the same old ways. Martino's developed system of articulation markings deals with that problem by providing a number of quite new symbols (fig. 36b). There are probably too many for a practical notation (players are not avid decoders at best), but a simpler system built along these lines could resolve many problems in articulation, allowing for a richer and more refined impasto of composed nuance. One should bear in mind that the articulatory possibilities are not necessarily available on all instruments.

I first used phonetic notation for attacks in my *High Flyer* (1969), for solo flute, to control details of the "air sounds" that are available from middle C to the E a tenth above. To the best of my knowledge these sounds were discovered and first put to musical use by David Gilbert. To produce air sounds the player blows either across the mouthpiece (the sounds that I have notated as square noteheads) or directly into the mouthpiece (triangular noteheads). Key slaps are notated as *X*'s, and the phonetic notation below the notes—"coffee, tea, etc."—contains the information about type of attack and mouth shape (see fig. 37).

Flute air sounds are particularly malleable and adaptable to phonetic control, both as to the consonant of the attack and vowel color. The verbal "nonsense" within which the directions are coded is not meant to be understood by a listener. One can make a simpler score, with less chance of performer error (and certainly less performer balkiness) by using speech sounds in conventional spellings, rather than the more precise but harder to decode international phonetic alphabet.

FIGURE 36*a*

PHONETIC REPRESENTATION	DESCRIPTION	SUGGESTED SYMBOLOGY	ALTERNATE SYMBOLOGY
1 ta, (ka/ga)	incisive, crisp attacks	•	ᴠ
1a pa, (ba/ma)		ᴜ	•
2 da, (na/la)	weak, delicate attacks	–	–
2a ha, (ja, wa)		h	h
3 (e/o), a	minimized, barely audible or non-existent attacks.	⌢	⌢

Note on pronunciation. "a" as in father, "j" as in yes, "e" as in bet.

FIGURE 36*b*

From this general classification the following list of modes of attack and decay has been constructed:
> is an operator which when applied to any one of the following symbols intensifies it by increasing its attack rate and dynamic (slightly).

SUGGESTED SYMBOLOGY	ALTERNATE SYMBOLOGY	PHONETIC REPRESENTATION AND DESCRIPTION	VISUAL REPRESENTATION
		tat: incisive, crisp attack with similarly closed decay	
		tap	
		tad	
		ta: crisp attack with open decay	
		pat	
		pap	
		pad	
		pa	
		dat	etc.
		dap	
		dad	
		da: weak attack, open decay	
		hat	
		hap	
		had	
		ha	
		at: minimized or inaudible attack, sharp decay	
		ap	
		ad	
		a: minimized or inaudible attack (legato), open decay.	

* As a former clarinetist I think I am correct in asserting that all starred symbols are appropriate to the clarinet.

Copyright © 1966 by Perspectives of New Music, Inc. Used by permission.

FIGURE 37

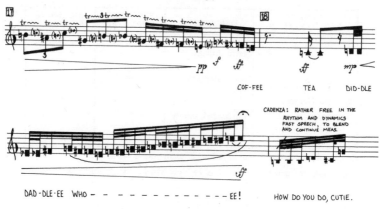

COF-FEE   TEA   DID-DLE

CADENZA: RATHER FREE IN THE RHYTHM AND DYNAMICS FAST SPEECH, TO BLEND AND CONTINUE MEAS.

DAD-DLE-EE WHO- — — — — — — — — — — — — EE!

HOW DO YOU DO, CUTIE.

My composition, *General Speech* (1968–1970), for solo trombone, exploits some phonemic possibilities of the trombone. Here the words are meant to hang on the edge of comprehensibility. Occasionally a word should be heard quite distinctly; more often the interest is in the play of timbre arising out of desemanticized speech, much in the manner of certain didjeridu players. The passage given in figure 38 is rather conservative from the point of view of manner of attack: /t/, /k/, and /d/ are part of standard technique; /th/ and /h/ are not particularly difficult. The /r/ sounds are harder to learn. Each of the attack consonants is significantly altered, of course, by their following vowels. (See the discussion on vowels in the section about spectral glide below.) The most difficult attack consonants for trombone are /p/ and /b/; these are difficult for other buzzed-lip instruments as well as for reed and edge-tone instruments.

FIGURE 38

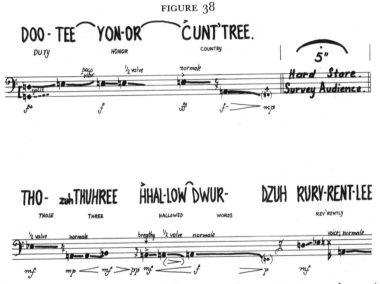

If a composer chose even a small group of consonants, such as /t/, /d/, /ch/(throaty) and /h/, all of which are available to most wind instruments, he would have at his disposal a wide range of accurately notatable attacks. Each consonant gives the wind player precise information about *what he should do*. It translates into motor terms with a minimum of cerebral intervention. It is accurate within a language community and it leads to repeatable results.

The essence of the attack of most acoustical musical instruments can be expressed in Schouten's term "prefix" with its implication of being anterior to and different in character from the continuation. Borrowing from the terminology of phonetics we may think of the beginnings of such tones as consonants, because the prefix is different from the ensuing part, which stands in the relation of vowel to consonant. It is not inappropriate, therefore, to think of many musical tones as syllables consisting of consonant and vowel (CV); CV syllables and CVC syllables provide a good initial framework for the analysis of microarticulation.

ATTACKS AND ELECTRONIC MUSIC SYNTHESIZERS

The attack (envelope) generators of electronic music synthesizers (Moog, Buchla, ARP, and others) provide means for controlling the attack time (onset) of a sound. They do not provide for such features as the individual growth of harmonics or the noise bands associated with consonants of the attack discussed above. Detail is sharply restricted, perhaps too restricted.

Timbre is not invariant under loudness change. Many acoustical instruments produce more higher partials as they are played louder, and composers have taken advantage of this for a long time, exercising a rough control over the spectral envelope through the careful use of dynamic markings. It ought to be possible to develop electronic circuitry that could mimic this important dimension of change; at the very least it is necessary to quickly update attack generators and their associated equipment to make a much broader range of attacks available.

One can argue that electronic music is an entirely new medium, with qualities different from acoustical instrument media, with different limits, different sounds, and different criteria. Varèse has shown in *Ecuatorial* that electronic instruments have their own integrity in an ensemble. Existing electronic synthesizer modules have a far greater range of attack length than any acoustical instrument. No other instrument can build up an attack so gradually over a time span of many seconds. We have learned the musical worth of these super-long attacks; already performers are becoming more adept at producing them on acoustical instruments.

Nor do I think that somehow electronic music should match the capabilities of all other instruments in every way, and I am far from desiring electronic sounds that merely imitate those of conventional instruments. Each instrument man has invented has its strengths and limitations. One composes to those strengths and limitations. My point is that music, most music, maybe not all music, needs detail, variety, nuance on the microlevel, and the technology of electronic music synthesizers is advanced enough to provide that variety.

CHANGE DURING THE PSEUDOSTEADY STATE

Many sounds cannot be described as having a steady state. They may be too brief or in a state of constant decay. Others, such as some electronically generated sounds and the sounds made by breath and bow, may have a more or less steady continuation after the attack. This portion of a sound, the vowel part following the consonant of the attack, is also thoroughly cinematic, exhibiting a multitude of small changes: periodic and aperiodic changes in amplitudes of spectral components; amplitude and frequency modulations; beats; changes in the rustle time of noise components. Expressed musically, some of these—vibrato, tremolo, vowel quality of the sound or richness of partial structure—can be controlled to some degree by instrumentalists and singers. The prevalence of vibrato shows an almost universal desire to keep the sound "alive," to shape it and keep it in motion.

WHAT VIBRATO IS

On the surface vibrato is a more or less regular oscillation around a pitch. Violin vibratos have a typical speed of about six oscillations per second, and the extent of the pitch change is approximately a quarter tone. Singers usually produce broader and somewhat slower vibratos, other instruments have typical speeds and extents. Tenney (1969) has pointed out that faster and narrower vibratos were more to his taste for the kinds of computer-generated sounds he was working with.

These regular changes are frequency modulations, but we do not ordinarily perceive vibrato in terms of pitch—we hear it as something having to do with the quality of the tone. If the vibrato is too slow or too wide (or sometimes too fast or too regular) for the particular pitch on the particular instrument in the particular musical context, then its pitch aspect may command our attention.

Fletcher and Sanders (1967) have confirmed many earlier studies of violin vibrato which showed that the modulations of frequency were accompanied by amplitude modulations at the vibrato rate. They also found that the variation in intensity level was different, very different, for different harmonics, and that the harmonic structure (spectral envelope) was constantly changing in time. Figure 39 shows the striking difference in spectral envelope for an E played on the A string between those moments when the vibrato cycle is at its highest pitch (maximum frequency level) and those moments when it is at its lowest (minimum frequency level). Vibrato tones that are unisons with open strings show especially large changes in harmonic structure during the vibrato cycle, and Fletcher has rightly drawn attention to the importance of the open strings to violin timbre as a whole, but all vibrato tones on the violin exhibit this change of spectrum during the vibrato cycle. Beauchamp (1972) has published a summary of the technical aspects of vibrato from an engineering point of view which takes fully into account the timbral/musical values.

Some instruments have a potential for superimposed vibrato; reed-instru-

FIGURE 39

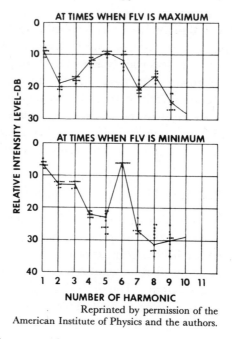

Reprinted by permission of the
American Institute of Physics and the authors.

ment players are able to produce a jaw vibrato and a diaphragm pulsing simultaneously; a trombonist who is sufficiently adept can use jaw, diaphragm, and slide for a multiple vibrato. These effects have been little explored but are potentially extremely useful.

Most attempts to explain the effect of vibrato, and its musical purpose, have pointed out that the broadening of pitch which is a consequence of the vibrato covers up small intonation errors that otherwise might be more noticeable. But there are two other important vibrato effects: (1) vibrato tones stand out, that is, they are attention getting, in textures consisting of vibrato and nonvibrato sounds; (2) vibrato may be used to control and express the microrhythm of the sound. This rhythmic aspect of vibrato, together with the possibility of composing other "pulsing" aspects of sound by means of electronic techniques, is likely to be exploited more fully. There is already a considerable body of "beat music" which takes advantage of the strengths and weaknesses of available electronic equipment.

BEATS, ESPECIALLY SLOW BEATS

If two tones differ only slightly in frequency and if their intensities are approximately of the same magnitude we can hear a periodic waxing and waning of loudness of a single pitch. This waxing and waning may be very slow, on the order of many seconds, if the pitches are very close together. As the two pitches are separated further the beat frequency increases. There are three stages in beat perception: (1) noticeable oscillations, waxing-waning; (2) intermittence, separate pounding pulsations; and (3) roughness without discernible pulsation.

Beats have been discussed in the literature chiefly as something to either minimize or eliminate, especially by theorists of consonance/dissonance bent upon creating systems with the largest number of beatless "pure" intervals. Beats have become musical material largely because electronic equipment allows for composer control of beat frequency over longer time spans than is practical for most acoustical instruments; however, I can offer at least two instances of acoustical instruments where beats are an important ingredient of the sound: the piano and the Indonesian gamelan. Many of the pitches of a piano are triple strung. The three strings, all struck by the same hammer, are tuned each to a slightly different pitch in such a way that the maximum difference is approximately 1.5 cents (Kirk, 1959). The piano tone is enlivened by the very slow waxing and waning produced by the running-phase relations between the three strings. We are not aware of the sound as a beat phenomenon, but if the three strings are tuned exactly together (or too far apart) the sound will have perceptible differences, usually felt as differences in quality.

All gamelan generate beats, in the manner of the gongs and other unpitched metal instruments of Western music; but the Balinese carry this farther. They build certain pitched metalophones in pairs, the "female" of the pair being pitched slightly lower than the "male." The shimmering quality, due to beats (Hood, 1966), which emerges when they are played together, is highly valued for its musical effect. This shimmer may be heard to advantage in *Music for the Balinese Shadow Play* (Nonesuch H-72037).

RUSTLE NOISE

Another element important to timbre in the time course of a sound is rustle noise, which, though aperiodic, is characterized by an average time between pulses. A Geiger counter, which is an excellent rustle-noise source, produces clicks which become faster as a radioactive source is approached. These random clicks can nevertheless be described in terms of a mean time interval between clicks, Schouten's rustle time.

Rattles of all sorts are differentiated by rustle times, determined chiefly by the fineness of the sand, seeds, or shot. Sizzle cymbals, drum snares, drum rolls, and string drums depend upon rustle time. Various weights and widths of bows and different types of hair and rosin make subtle differences in string sound, attributable in part to rustle time, and the concept need be extended only slightly to accommodate flutter-tonguing, growls of brasses and wood-winds, resonated vocal fry (woodwinds), and the eructation sounds obtainable on some wind instruments. Rustle noise can easily be generated electronically by using the amplitude peaks of white noise to trigger a pulse generator.

The least explored aspect of rustle noise is its easily controlled variability along an *accelerando-ritardando* scale. Smooth changes of rustle time are possible on a few acoustical instruments. Instruments should be explored for further possibilities of variation of rustle time because this apparently minor aspect of musical sounds has a disproportionately large importance for higher levels—textures, ensemble timbres, contrasts between musical events.

SPECTRAL GLIDE

Buzzed-lip instruments, especially those with large mouthpieces, have a potential for extensive modification of the vowel quality of a tone, as in the solo part from my *Ricercar à 5,* for trombone (fig. 40). This passage allows for a comparison between blown and sung vowels. One has to listen closely to detect the two different manners of tone production.

FIGURE 40

Reprinted by permission of Okra Music Company.

Woodwinds, except for flute in the air-sound mode, have a smaller range of vowel quality, and strings have the smallest, although all strings are able to produce a distinction between the nasal sound of *sul ponticello* and normal bow position. The musical possibilities of *poco a poco sul ponticello* and *poco a poco sul tasto* in a single, slowly drawn bow have been little explored, and probably a considerable range can be brought into practical use, particularly on cello and contrabass.

Tuba and trombone players easily produce common vowel sounds, and with practice are able to change vowel quality either while maintaining a steady pitch, as in the passage from *Ricercar à 5* (fig. 40) or concurrently with a change of pitch, as in the word "create" in the passage from my *General Speech* (fig. 41). If the vowel change happens quickly we may not notice the spectral glide as such. In the two passages in figures 42 and 43 each uses the change from /ee/ to /oo/ in the word "you." The /ee-oo/ of "were *you* to do so" is much less discernible as a spectral glide than the /ee-oo/ of "Today marks my final roll call with *you.*" Whether perceived specifically as a spectral glide or as part of an overall "quality," these one-way, nonoscillating changes of spectral envelope are of cardinal importance for shaping individual tones.

FIGURE 41

FIGURE 42

FIGURE 43

Spectral glides can also be generated by add-on devices, such as the familiar wa-wa mute, as in the passage from *Punkte*, 1952/1962 (fig. 44), where Stockhausen has notated the glide pattern in terms of transitions (𝗼𝗼 or 𝗼𝗼) between open and closed. Now that the effect has become generalized by the invention of a wa-wa pedal, electronically produced spectral glides have been appearing increasingly (to the point of nausea) in television commerical messages and other expressions of popular culture.

With some effort a composer can avoid the grossness of the electronic effect, as A. Wayne Slawson has done in his *Wishful Thinking about Winter* (Decca DL 710180), a computer-generated setting of a Japanese Haiku poem which grows out of speechlike sounds and which uses an exceptionally wide range of spectral glide rates. Loren Rush started a computer-sound–generation project in 1967 which could uncover unsuspected possibilities. Starting with a bassoon-like sound and a sound approximating that of a bass clarinet, he explored the question of what the quality would be of a sound halfway between and devised a program to produce a smooth transition between the two stable spectra. It is not hard to imagine compositional strategies based upon a hierarchy of fixed timbre points connected by means of spectral glides whose tempi could be further organized on some rhythmic plan.

Clearly the speed of the spectral glide is an important parameter. If a glide

FIGURE 44

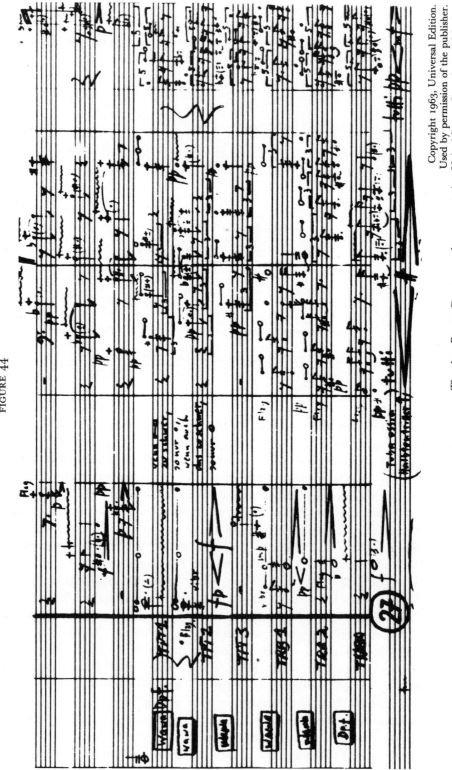

oscillates between two more or less fixed states at rates approaching six per second then we will perceive some sort of vibrato/or tremolo. If the glide is too slow, then any change of spectral envelope is likely to go unnoticed; or we may become aware of individual frequency components and hear it in terms of pitch. Our pitch-centered tradition, prizing invariance of melodic motive under transposition, has led us to deemphasize the timbre component of pitch, but, as investigators beginning with Helmholtz and Stumpf have often pointed out, even sine tones have a vowel quality, ranging from /oo/ to /ee/, depending upon position in the frequency scale. If the pitch parameter of music is not organized in discrete scale steps (and much electronic music is not), and if unidirectional frequency modulations are of the right speed, we have a situation where this kind of quality change can be a function of change of pitch. Anyone working with vowel-like spectral glides will need to take this into consideration. Not all vowel-like contrasts are available in all pitch registers.

GRAIN

This chapter has been concerned with short-time events of 10-200 milliseconds (attack lengths); six or seven per second (vibrato); beats, from the fast beating of partials to a beat in several seconds; and spectral glides of various speeds. Some of the spectral glides between vowel-like and other speechlike sounds can be as short as 15 or 20 milliseconds. While the attention I have paid to these short-time events implies that they have practical value for composers and performers, it must also be obvious that their *musical* possibilities depend upon the compositional and performing context in which they occur. The time intervals involved are so short that a critical reader might feel that such a mass of tiny details has no great musical relevance; he might point out that a performing space is likely to have an appreciable reverberation time, and that it is of small use to talk about events on the millisecond level, which will be smeared anyhow. Or a critic might say that even if one accepts as true the physical existence of all the short-time details described, those details may not be relevant, and that what is important is not the grain of sounds but the grain of music, even the grain of a particular composition. Nevertheless music is full of short-time events, and one need not go to the attack level to find times on the order of 100 milliseconds. To relate milliseconds to musical time values and to illustrate short times in common use, consider some typical note values given in figure 45: the diagram on the left has the quarter note at MM 60, or one per second, and in the diagram on the right it has a value twice as fast, MM 120.

Some of the attack times measured by Melka and others were longer than 125 milliseconds! Electronic music has many passages of more or less discrete

FIGURE 45

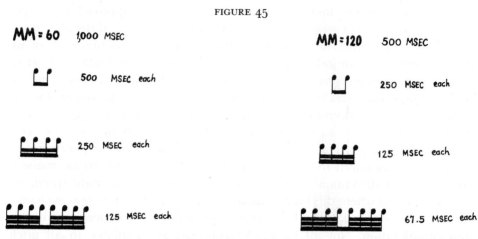

notes moving at speeds much faster than 125 milliseconds. The neighborhood of thirteen or fourteen per second is widely quoted as the fastest speed at which a pianist (or a typist) is able to articulate individual pitches, and it is characteristic for other acoustical instruments as well, although one would expect that short groups of notes that lie well under the hand could be performed considerably faster. At a tempo of ♩ = MM 60 we can easily find the millisecond values for division of the quarter note into 5, 6, 7, and more parts, set out in round numbers as follows:

Notes per second	Milliseconds each
5	200
6	166
7	143
8	125
9	111
10	100
11	91
12	83
13	77
14	71
15	67
16	62
18	56
20	50

Another way to demonstrate the importance of these short times is to imagine an ensemble attack as in figure 46, a chord for flute, oboe harp, clarinet, and horn. The composer may have been hoping for a simultaneous attack; as written, the synchronization is likely to be rough. If maximum simultaneity is the consideration, the composer can alter the dynamics, change the voicing, or even use other instruments, but he must be aware (not in an intellectual, let alone numerical, way; he must have stored memories of how each instrument sounds at any dynamic and pitch level in many typical musical situations) that the flute playing low C will produce a rather long attack and the harp a much shorter one with much less frequency dependence than the flute.

FIGURE 46

The technical problems are very different for a composer of electronic or computer music than for a composer using acoustical instruments. His generating devices are capable of speeds far faster than the thirteen or fourteen per second of human performers, and, at least with computer synthesis, he can compose details as fast as desirable. Electronic synthesizers allow for hearing sounds at the same time as they are produced, but computer synthesis at present does not have this on-line capability, except for a few systems, such as Max Matthews's Groove. Moreover, computer programs are written in a digital code, and this means that all musical events must somehow be expressed in discrete numbers. That is why almost every discussion of computer music includes a plea for more psychoacoustic information.

Nor can any consideration of microtime leave out the performing and listening space. The hall, and especially its reverberation, must be considered. With the introduction of new instruments, electronic enhancement and reinforcement, new kinds of music, we can no longer use a generalization like "good acoustics." Most definitions of "good acoustics" are in terms of spaces where opera and symphony take place. There is a growing tendency to view those genres as belonging to history and their specialized concert halls and opera houses as expensive entombments. Now that it has been (re)discovered that the space may be composed, there is a need for entirely new and flexible conceptions of performing space. Different musics need different spaces. There are outdoor musics and indoor musics. Their differences are not trivial, and the performing space has always been integral to the musical conception.

REVERBERATION FOR BATS AND MEN

Most musics of interest today are performed indoors. Reverberation imposes certain limits and constraints upon the time course of single and multiple timbres and their succession.

The strange faculty humans possess for listening to significant beginnings embedded in the mass of echoes and reverberations of an ordinary concert hall is just beginning to be investigated. We take that faculty so for granted that it is hard for us even to notice its existence; but if we look at it from another point of view—that of animal echolocation—we can gain some insight into how it might affect musical microtime and short-time timbre changes.

D. Griffin asks, "Why is that blind men cannot echolocate as well as bats and dolphins? Where bats catch fruit flies at rates of several per minute, blind men cannot safely drive automobiles, let alone fly airplanes to catch birds—which of course, would be a directly analogous performance. Is there something qualitatively different about bat brains that allows discrimination of echoes imperceptible to much larger and more complex human brains?" (1968:156).

He goes on to say that there are no observed basic neurological differences, but that the middle-ear muscles of bats serve to reduce auditory sensitivity during the emission of each pulse of orientation sound and that these muscles relax fast enough—a few milliseconds—for good sensitivity to echoes.

The sound experience of human beings is fundamentally different. There is a subjective reduction in loudness of echoes after a much louder sound, and

it is not altogether clear how this reduction in subjective loudness of echoes is achieved in the human brain, or even at what neuro-anatomical level it occurs. But the information of vital importance to echolocation is obviously contained in the same time span as that during which this suppression of echoes occurs. Perhaps this echo suppression is absent in echolocating animals, and if so, this might explain their superior ability to react to echoes following within milliseconds after a loud orientation sound. These compounded speculations can be useful only in so far as they stimulate new inquiry, and preferably direct experimentation. But they lead to an inverse question, why have human beings acquired the echo-suppression mechanism in the first place? Could it be helpful, if not essential, for the discrimination of speech, especially indoors or in situations where multiple echoes confuse the waveforms of impinging vocal signals that must be analyzed in order to extract meaningful information from messages of another member of the same species? [Ibid.:160–162]

G. Békésy has made some preliminary investigations of backward masking which have direct bearing on reverberation and musical detail. One can find the forward masking referred to by Griffin in one's ordinary experience with music, but backward masking, backward inhibition, seems intuitively implausible.

FIGURE 47. (A) A central loudspeaker (No. 1) and a ring of loudspeakers (Nos. 2) are used in combination, with an interval of 60 milliseconds between arrival of the sound of speakers 2 and 1, or vice versa. (B) If the central speaker is presented second, this inhibits to a very large degree the sound of the speakers in the ring, and the extension of the sound image around the central speaker is reduced. If the speakers in the ring are presented second, they almost completely inhibit the sound of the central speaker and we have an apparent extension of a sound image, represented by the shaded area in B. (C) Diagram showing that the stimulus presented second accounts for a longer time interval in the perception than the stimulus presented first does.

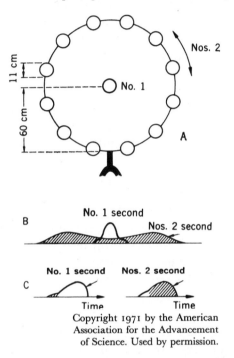

Inclined as we are to believe that events in the brain occur in an orderly sequence, we have always been somewhat amazed by a phenomenon is which one stimulus can be completely erased by a second stimulus arriving much later (about 60 milliseconds later). But if we assume that every stimulus starts a process in the brain which lasts perhaps 200 milliseconds, we can make backward inhibition acceptable if we further suppose that this process can be inhibited at any moment during the 200-millisecond interval by the onset phenomenon of the second stimulus. With a delay greater than 200 milliseconds no inhibition occurs and both stimuli are recognized, separately. [1971:530]

Békésy used 35-millisecond bursts of tone and noise with varying time intervals between sound pairs. In order to simulate room acoustics he constructed a ring of loudspeakers and a single center loudspeaker, as in figure 47. An observer standing 2.5 meters outside and in front of the circle was asked to make a graph of the local sound density of the center speaker. In this experimental situation the sound that comes 60 milliseconds later "almost completely inhibits the earlier sound" (ibid.:532). Depending upon whether the first sound comes from the ring or the center the extension of the sound image is either enhanced or reduced.

Békésy correlates this 60-millisecond optimum delay time with direct and reflected paths, calculated from the dimensions of several well-known concert halls. His investigations may also throw light on such matters as simultaneous attacks—why we hear a single *tutti* sound although individual musicians are early or late; and the blurring of timbral distinctiveness and contrast at fast speeds. Additional work on backward masking in a performing space may lead to a more complete explanation of processes underlying our ability to "hear beginnings."

Investigations such as these (Békésy enumerates five types of auditory inhibition, each different for a continuing and a transient tone!) dramatize the complexity of our mental processing and remind us to be cautious with blanket statements about these largely unexplored regions of microtime.

AESTHETIC INTERLUDE: HIERARCHIES AND SYSTEMS

If we wish to think of music in terms of grain we might attempt to discover the smallest quantum. But music is notoriously hierarchical, and the size of the grain would have to vary from finer than a single sound to large groupings of notes, depending upon composed relationships. Listening to music is an active heirarchic process; therefore, what we hear (understand) will depend both upon the composed relationships and the grain of our listening.

It is not especially fashionable these days to draw attention to hierarchical structure in any of the arts. Struggles to articulate new artistic conceptions seem also to demand extensive rejections, renunciations, and reformulations of inherited ideologies; and in an era of news and newsworthy art, such as ours, ideological literature proliferates. The public occasionally hears an unfamiliar work whose structure is not obviously hierarchical in the manner of the eighteenth and nineteenth centuries; the journals present reams (some of it mere sportswriting) about mosaics, serial organizations, indeterminacy, and so on. Regardless of the newsworthiness of the theory of the day, the hierarchical nature of music is not dependent upon aesthetic formulations. Hierarchy will function in any music or in any succession of environmental sounds that attracts our interest, because it is built into the way we perceive. H. A. Simon, writing about hierarchy in physical and biological systems, says: "If there are important systems in the world that are complex without being heirarchic, they may to a considerable extent escape our observation and our understanding. Analysis of their behaviour would involve such detailed knowledge and calculation of the interactions of their elementary parts that it would be beyond our capacities of memory or computation" (1969:108).

Wladimir Weidlé's *Biology of Art* presents related notions in a closely drawn biological metaphor: "The tissue of a work of art seems alive because it closely imitates—as closely as possible—the cellular tissues of organisms. And this is how it does so: it is entirely composed of units of brief duration or of reduced size which, in turn, imitate the internal structure of living cells" (1957:11).

I willingly accept the implications of that metaphor as a basis for musical analysis. If someone protests that the relation between a work of music and a living organism can never be more than an analogy, then my reply is that at least the analogy between music and biological forms and processes has a reasonable basis: art is something peculiar to man, and men are living beings composed of living cells.

Terms such as "alive" and "dead" may be no more than convenient tags; if an art work is alive it is alive in some way which is probably undemonstrable. Weidlé is careful to say "seems," not "is." Even foregoing "aliveness" one can be quite specific about the extensive parallelism of structure (for that is what makes Weidlé's metaphor interesting) between music and organisms. Stated bluntly, a musical composition is a system in the same way that an organic entity is a system. This is meant as a simple, nonmetaphysical statement. As the biologist Paul Weiss put it: "*The principle of hierarchic order* in living nature reveals itself as a demonstrable fact, regardless of the philosophical connotations it may carry," and "the necessity becomes compelling to accept organic entities as *systems* subject to network dynamics in the sense of modern systems theory." (1969:4).

Systems theory, or general systems theory, is a relatively new discipline which seeks to discover principles applying to all systems, whether biological, physical, sociological, or symbolic. It seeks to place the analysis of wholes and "wholeness" on a logical-mathematical base. Weiss defines a system pragmatically as

a rather circumscribed complex of relatively bounded phenomena, which, within those bounds, retains a relatively stationary pattern of structure in space or of sequential configuration in time in spite of a high degree of variability in the details of distribution and inter-relations among its constituent units of lower order. Not only does the system maintain its configuration and integral operation in an essentially constant environment, but it responds to alterations of the environment by an adaptive redirection of its componental processes in such a manner as to counter the external change in the direction of optimum preservation of its systemic integrity. [Ibid.:11–12]

This is a definition of an "open" system, that is, one that may approach a steady state (maintain its configuration and processes) although irreversible processes are going on within the system. My friends recognize me over the years even though individual cells in my body are constantly being replaced, and I recognize a musical composition when individual musicians in the performing group are different at each performance. Closed systems are more characteristic of inanimate physical processes in nature.

At the risk of belaboring the obvious let me sketch an overview of how this definition of an open system can be related to music:

"A rather circumscribed complex of relatively bounded phenomena" could be a whole music, a complete composition, a movement of a composition, or a part of a movement. Within the bounds described it "retains a relatively stationary pattern of . . . sequential configuration in time"—think of a movement of music which allows the player to apply certain ornaments at his discretion, or of two performances of the same composition, entailing a multitude of differences in articulation, tone production, and phrasing—"in spite of a high degree of variability in the details of distribution and inter-relations among its constituent units of lower order."

It is these "constituent units of lower order" which are so interesting. It is trivial merely to draw distinctions between the overall organization of a piece of music and nuances of "interpretation." One wants to know more about the level or levels between those extremes. The concepts of Heinrich Schenker apply with some cogency to certain dimensions of tonal music, but it does not necessarily follow that they can be applied fruitfully to either rhythm or timbre, even though they rest upon hierarchic and system principles. A search for appropriate "constituent units of lower order" which might be of value to timbre analysis ought to be wide ranging enough to apply to all music.

To go back to Weiss's definition—his final sentence refers to how a system maintains its configuration in a constant environment and how it counters alterations in the environment "in the direction of optimum preservation of its systemic integrity." It is well known that hall conditions, such as the amount of reverberation, number of persons in the audience—all elements involved in the feedback processes of performance—require precisely these "countering" alterations from the performer in the direction of "optimum preservation of the systemic integrity" of the music.

TIME GROUPINGS IN MUSIC

Music is preeminently a temporal system, and there may be limits to what we can perceive or what we do perceive in various musical situations. Are there time constants? Limits to augmentation and diminution? Must a klangfarbenmelodie be relatively slow?—slower than a melody in a homogeneous timbre? If limitations upon fastest and slowest can be uncovered, to what extent can they be influenced by aspects of music which are under the control of either composer or performer?

FIGURE 48

*Diuersarum uolucrium voces
notis musicis expreſsæ*

FIGURE 49

These questions open musical vistas (battlefields would be a better word) where beliefs are strongly asserted and analysis seldom enters, so I would like first to test some ideas about temporal limits with nonmusical materials, the songs of birds and the songs of whales.

When Athanasius Kircher (1650) put the crow of the common barnyard cock into musical notation he specified it somewhat more precisely and in more detail than "cock-a-doodle-doo" (fig. 48). It is easy to hear the rhythm and pitch relationships of typical cockcrows from his notation. But if a cockcrow is carefully transcribed from a slowed tape recording, using transcription techniques such as those developed by Bartok and Kodaly for folk music (Szöke et al., 1969), then much more detail is revealed than we normally perceive (fig. 49).

C. Greenewalt compared time discrimination for birds and men, using 50 milliseconds (twenty per second) for the greatest number of discrete sounds humans can perceive per second, after Martin Joos's suggestion, based largely on the speed of phoneme discrimination—hardly more than a ballpark estimate. From measurements of the small amount of deviation in repetitions of phrase segments in several bird songs, Greenewalt concluded that "There is

FIGURE 50

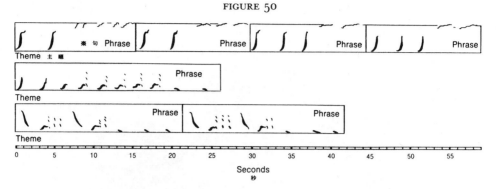

From CRM Books, *Songs of the Humpback Whale,* © 1970 by Communications Research Machines, Inc. Reproduced by permission.

a strong presumption that birds hear *as such* the rapid modulations so characteristic of their songs, and that the information content even in their relatively simple songs must be enormous" (1968:142).

At the other end of the time scale are the very long songs of the humpback whale, recorded by Roger Payne (1970). The shortest song recorded ran six minutes and the longest more than thirty minutes. It is very difficult for humans to hear the patterning in these songs because our attention is engaged by details of individual phrase elements such as represented in Payne's spectrograph-based notation (fig. 50). The time scale of Payne's "phrases" and "themes" is too large for quick comprehension; it would be interesting to see the effect of practiced listening.

At very fast speeds we abstract and simplify, constructing a pattern that has some correlation with elements of the signal. Notice that in the Szöke cock transcription (fig. 49) there are accented notes, long notes, and loud notes to provide cues for this construction.

The whale song presents a different musical problem, more like learning to understand and appreciate an extended musical composition, one with few clearly defined motives but long and diffuse phrases. Anyone who has tried to grasp rhythm patterns is likely to have discovered that complex rhythms become easier to hear when the tempo is faster, often to the point of sounding trite.

It does appear that, within broad and by no means fixed limits, there are "music-sized bites," and that when the signal contains too many discrete elements we read groupings of elements as units. Babbitt's investigations of pitch sequences and pitch succession, as reported in a lecture at the University of Illinois, support this view. He found that as one reduces the speed of very fast pitch sequences that are mere blurs at the fastest speeds, "every single pitch that can be recognized at every stage of the reduction of speed is the extremum of a contour, the highest or lowest note of a succession in the same direction" (1965: tape recording).

What about timbre? Do we simplify and abstract, gain and lose timbral detail in music depending upon the time scale? I explained in the first chapter how the effect of constancy of timbre under pitch change could be understood in terms of "overlooking" in our perceptual processing: less contrast, more constancy. Fast melodic motion—and much music, including eighteenth- and nineteenth-century music, where sixteenth notes often are moving at rates of 200 milliseconds or faster—can enhance this "overlooking," because attention is likely to be focused primarily upon pitch.

Pitch complicates the situation enormously, and it may not be possible ever to create conditions (insofar as they are musical conditions) where pitch is not a factor, and an important one. One can limit pitch influences, however, at least experimentally. One can assemble a tape of noises of various sorts, and, by splicing or other means, make sequences of very fast timbre contrasts. In general one finds that the faster the speed the less the contrast. One can improve the contrast, perhaps even optimize it for the particular configuration, by shifting and arranging the tape segments to taste; nevertheless, the general trend is toward less contrast and less distinctiveness at faster speeds. Some of these speeds are not even very fast compared with common speeds of pitch succession. Timbre contrasts that may be strikingly well defined at speeds of one per second can lose distinctiveness at speeds of two per second.

These tape-segment experiments are by no means the last word. Contrasts in music, including timbre contrasts, are not often found to be moving at some fixed unit speed, and if one constructed sequences where the tape segments were of different lengths, and where the attack aspects were taken into account, one could compose (articulate) more contrast. If one does this, though, he will find that he has made groups of some sort, units larger than a single sound. There is no way to make such composed groupings of timbre contrasts which does not also involve rhythm, dynamics, and pitch. If pitch is excluded as much as possible we find that loudness relationships between the segments and differences in length are a powerful influence upon the effective contrast, and that as speed is increased the contrast between the single-segment details tends to be lost, supplanted by contrast among groups. One would like to know a great deal more about the formation of these groupings in musical situations, not to discover some absolute limen of successiveness or to derive "rules and regulations" for fast music, but to understand the working of units and subunits of timbre which might be composable.

TRADITIONAL USES OF TIMBRE CONTRAST

The musical styles of seventeenth-, eighteenth-, and nineteenth-century music placed a heavy emphasis upon the carrier aspect of timbre. Strong timbre contrasts usually involved changes of instrumentation, and such changes could be very slow, even to remaining fixed for a movement or a section. Composers such as Mozart, Haydn, and Beethoven often used faster changes of instrumentation, down to the phrase and motive. In view of how often timbre change is integrated to units no smaller than phrase or motive (and this is true of much twentieth-century music, too, including most of Webern's music), how striking are the measures from Berlioz's "March to the Scaffold" from *Symphonie Fantastique* (fig. 51). The meter is 4/4, MM 72 = ♩, so the speed of a quarter note is considerably faster than two per second. The contrasts are mostly among choirs of the orchestra, including percussion(!) on the final quarter of measure 111, but Berlioz also exploits the differences between bowed and plucked strings. The musical sense of this passage is essentially timbral; all other musical dimensions have a secondary role. The conception is daring. If one is looking for firsts, mark this passage as the first bit of klangfarbenmelodie of modern times in the West.

Debussy sometimes juxtaposed instruments quite closely, as illustrated in the passage from *Jeux* in figure 20; and he always composed instrumental contrasts with a profound sensitivity to the nuances of those changes. A passage toward the end of *"Fêtes,"* from the *Nocturnes* for orchestra (1893–1899) (figs. 52, 53) shows that artistic conceptions concerning timbre were well established before the turn of the century. The phrase at *Un peu retenu* (fig. 52) (observe the tuba!) is shortened to its first three notes, which are passed between woodwinds until rehearsal number 21 (fig. 53). The *cuivrez* sound of muted horns interrupts, and a dissolution follows (fig. 53). The changes of timbre are essential to the effectiveness of the dissolution, and are involved in the striking contrast at the *a tempo.*

In all his works Mahler sought out extreme contrasts of all kinds, and it is no surprise to find the timbral dimension included. There are some passages, particularly in the later works, where the substantive contrast is expressed timbrally, for example, between rehearsal numbers 19 and 20 of *Song of the Earth* (1910), where some of the contrasts between instrument groups approach the ensemble timbres discussed in the second chapter. The three measures in figure 54, where contrasts are only lightly bound to phrase and motive, are nevertheless supported and given continuity by the ongoing pedal A of the contrabass.

The final measures of the "Nachtmusik" from Mahler's *Seventh Symphony* exhibit a similar conception and technique, except for a few bars at measure 221 where there are some unusual changes of instrument from bar to bar (fig.

FIGURE 51

From Edition Eulenburg Pocket Score no. 422. Reprinted by permission of C. F. Peters Corporation, sole selling agents for the United States.

FIGURE 52

FIGURE 53

FIGURE 54

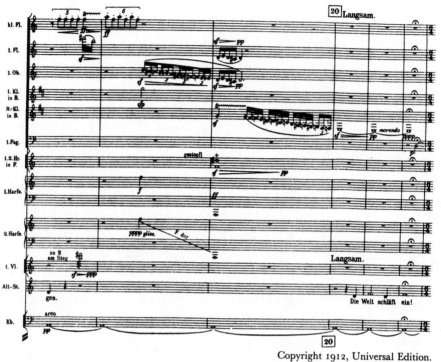

FIGURE 55

55). Guitar joins flute at 221; second flute joins guitar at 222; clarinet joins guitar at 223; the sequence is unified by the guitar timbre, of course, and the composer has obviously been careful not to have the passage break into individual bits. Mahler's concern with contrast *and* continuity here is reminiscent of Webern's setting of the opening measure of the first of his *Five Pieces for Chamber Orchestra*, Opus 10 (see fig. 4).

Composers are no longer so cautious about the continuity aspect of timbre change. Rather fast rates have become commonplace with the advent of electronic and computer-generated sounds. Significant changes, contrasts bearing upon musical relationships, are harder to find because they are harder to make. If performers are involved there is a communication problem: What cues—visual, auditory, or gestural—can help him to judge how the sound he makes should fit into the overall pattern?

Morton Feldman has devised a notation for specifying changes of instrumental grouping which is practical enough for orchestral musicians to read. The score is essentially a graph of grouping and change, with a time quantum of MM 80 = □ (fig. 56). The number in the box tells the player how many notes to play. A player who must play seven notes at the tempo of MM 80 must move quickly. It is easy to see that while the contrasts may be very fast, the means the composer uses for controlling grouping is limited to change of instrument. Timbre contrast (continuity too) can be rather well controlled, in spite of the limitations—no notation of dynamics, articulations, or pitch. The composer specifies that the dynamics throughout are "to be kept very low," and that "the attack of each sound should never be accented." Except for a few sections where low pitches are specified, all "sounds chosen are to be played in the high registers of the instrument." A few plucked and struck instruments are exempted.

Since the pitch is free within a certain register, the players, if so inclined, can concentrate on tuning the timbre. Study of several performances of "out of last pieces" (fig. 56) might reveal more about units and subunits of timbre larger than a single sound.

Each of the examples I have discussed can be analyzed for contrast or for continuity of timbre. The two functions of timbre can never be completely separated although there are musical situations where timbre functions more as carrier and others where the interest is in timbral objects.

These two functions can be uncovered at every level of temporal magnitude: a single sound has parts; the relation between two notes of different timbre will have elements of contrast and continuity; two or more different sounds in succession can group into a unit if the speed is fast enough and if other conditions, such as pitch range, rhythm, dynamics, permit.

Contrast is speed dependent; until the twentieth century timbre contrasts were slow, several minutes to a few seconds. At faster speeds discrete change and continuity have some special characteristics: when too many contrasts are

FIGURE 56

... OUT OF 'LAST PIECES'

Morton Feldman

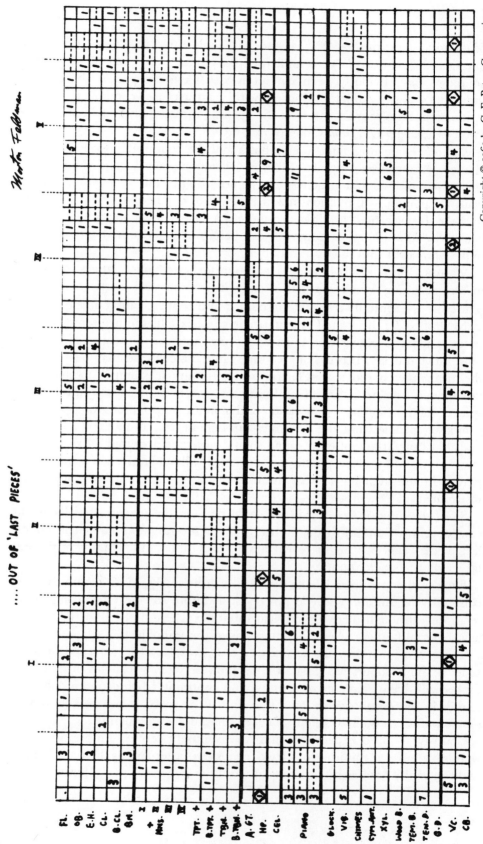

compressed into too short a time span the individual contrasts lose distinctive-ness and are perceived either as a blur (continuity) or as more or less well-defined single percepts (grouping). Composers who are interested in fast moving—even moderately fast—klangfarbenmelodie will need to pay as close attention to the effects of tempo, rhythm, dynamics, accent, and pitch upon timbral succession as composers concerned chiefly with pitch have paid to skips, steps, motives, phrases, and their articulation.

4

Drones

Drones have been few in Western concert music. Our tradition has been so oriented toward pitch, pitch change, and pitch relationships that probably most listeners are familiar only with the drones of bagpipe music and the occasional pedal points of seventeenth-, eighteenth-, and nineteenth-century music. Pedal points are understood in terms of harmonic function; bagpipe drones are suffered as a noisy adjunct to a loud national music.

Interest in drones and in drone-oriented music may be due, in part, to concern with timbral change and timbral organization, for it is obvious that if pitch is constant and unchanging we are free to attend to other dimensions—timbre, texture, rhythm, and so on.

There can be little doubt that drones are primitive, if by primitive we mean ancient in man's music making. It is easy to categorize contemporary drone music as merely a manifestation of current trends toward primitive and/or minimalist art. Some drone music is no doubt minimalist, purposefully impoverished, primitive in the sense that it lacks depth of structure. But such impoverishment is not a necessary condition. The manner in which a composer of drone music uses timbre, rhythm, and texture will determine its complexity of structure.

Wagner's Prelude to *Rheingold* (1854) is the only well-known drone piece in the concert repertory. Within the time scale of the complete Ring cycle it is hardly more than the first chord of a vast progression; nevertheless, it is long in real time, long enough for listeners to feel the absence of root movement or chord progression.

It is a true drone piece in that the lower instruments sustain an Eb throughout. Eight contrabasses are reinforced by a contrabass trombone and a contrabass tuba. Orchestral forces are deployed for maximum blend: the dynamic marking is *piano* throughout (the Prelude is usually played too loudly); the range is mostly low; eight horns enhance the homogeneity of the sound; there are no sharply profiled motives. Notice, for example, how the arpeggiated triad, symbolizing the Rhine, is imitated by the eight horns in such a way that the effect is thoroughly textural, more sound than imitative entries (fig. 57).

The wavy melodic movement rising later through the strings, more figuration than melody, is the chief source of forward motion and climax; and, although the introduction of faster notes gives some feeling of sectional division, the general effect of the Prelude is flowing and seamless, and its form a crescendo of texture.

Loren Rush's *Hard Music* (1970), for three amplified pianos, treats a traditional instrument in a new way, exploiting the differences in spectrum obtainable from the harder touches. Each performer plays a fast repeated low D (just above cello C). There is no melodic figuration. Thus the surface of the music consists of fast composite rhythms formed by the ensemble of percussive attacks and the interplay of the partials. The dynamic marking is *fortississimo* throughout, and the sound breaks apart in such a way that some of the partials, especially the fifth through the ninth, are easily heard as individual pitches. At especially loud places in the music the clangorous higher partials take over, and the effect is somewhat like distant bells. The content of the music is sound. The individual attacks are so closely spaced that musical changes are felt more as continuous than discrete, and one follows those changes as one watches clouds forming and reforming.

Folke Rabe's 1968 composition, *Was??* (Wergo 60047) is an entirely electronic drone composition, where the "cloud formation" structure is fully exploited. There are several central drone pitches involved, although the time scale of the composition is such that pitch changes are hardly noticed as significant change. There are no obvious discrete changes of any kind, and no sharply defined sections. The music makes much of the composite beat-fre-quency patterns mentioned in the last chapter, and—an elegant touch—the micropitch changes necessary to the production of beats are worked into a larger pattern of pitch relations. Spectral components are manipulated in such a way that a sound presents itself at certain moments as a unitary timbre and at others as individual pitch elements detached from a background. All possible timbral dimensions are manipulated: spectral envelope, including harmonic and inharmonic partials; time envelope phenomena, such as beats and tremolo; micropitch changes, both fast and slow. Transformations between pitch (with timbre)→chord, chord→"a sound," "a sound"→pitch (with timbre) abound.

I used another approach, emphasizing discrete timbre contrasts, in my composition for soprano, chorus, and tape, *Down at Piraeus*. The sound sources were mostly recorded instrumental sounds. One group was characterized by sharp attacks: electric guitar, toy zither played with wooden mallets, pizzicato

FIGURE 57

contrabass. Another group of buzzed-lip sounds, trombone and didjeridu, had somewhat longer attacks, and, for some sections of the composition, striking spectral glides. Very weak attacks combined with obvious spectral glides were produced by an ordinary pitch pipe played while altering the shape of the vocal cavity.

All of these source sounds underwent further processing: octave transposition, with appropriate equalization; band-pass filtering, to clarify individual sounds before mixing; modulation (pitch-pipe sounds were passed through a balanced modulator to create broad side bands); gating, to superimpose new attacks, especially upon pitch-pipe sounds, which were made to sound much like a jaw harp.

The central drone pitch was G♯ above middle C, but octaves from contrabass to piccolo range were also used. Change of octave was often involved in articulating sections, and for some of the denser textures five or six octaves were employed. Contrasts of the continuous variety were mainly of detail—pitch pipe, trombone, didjeridu; discrete contrasts articulated phrases and larger units, which were often quite differentiated by range, total band width, thickness and thinness of texture, and rhythmic organization.

VOCAL DRONES AND OTHER MOUTH MUSIC

Changes of timbre in drone singing, for example, like that of the Tibetan Lamaist chant discussed in chapter two, are controlled largely by the sequence of consonants and vowels. The texts of those Tibetan chants contain many interpolated syllables. According to Peter Crossley-Holland (1970) the purpose of the interpolations is to render the text meaningless to noninitiates. To the extent that the syllables are incomprehensible they are music, occasions for timbral nuance on a drone tone—timbre words—and the musical progression is ordered and regulated by their sequence. When semantic constraints are relaxed syllables and phonetic sequences may be composed entirely for their musical sound properties.

The single syllable *om* is the notation of such an ordered progression of timbres, and in Brahmanism it is anything but meaningless. As explained by H. Zimmer, the five final verses of the *Māṇḍūkya Upaniṣad*

bring the analysis of the . . . states of the Self into connection with the syllable OM, which, as made known at the beginning, is identical with the Self. In Sanskrit the vowel /o/ is constitutionally a dipthong, compounded of /a/ plus /u/; hence OM can also be written AUM. We read consequently, in the text: "This identical Atman, or Self, in the realm of sounds is the syllable OM, the above-described four portions of the Self being identical with the components of the syllable, and the components of the syllable being identical with the four portions of the Self. The components of the syllable are A,U,M." [The fourth component is silence.] [1956:376]

When used in secret and sacred ceremonial songs the sounds of speech are inseparable from the musical sound as a whole. This notion is not easy for us to grasp. In an oral tradition the sharp division between the categories of speech and music is often blurred, because of the way in which the ceremonial sounds—secret words, meaningless syllables, individual vowels and consonants, echoic and imitative sounds—are built into the music to form one single whole. C. J. Ellis, who has studied the music of the Aranda-speaking people of central Australia, makes the point: "In such a verse [i.e., a musically reshaped group of words expressing an idea or describing an action] the syllables of the component words have been changed so completely by their musical setting that the verse cannot be recited; hence each verse must be memorized in its chanted form" (1964:24).

One cannot doubt that the consonants and vowels of speech sounds have important effects upon song. These effects may be enhanced when nonlexical syllables predominate; they are surely enhanced when pitch is restricted to either a few notes or to a single note with microtonal inflections.

Probably this kind of music is very old. It appears to be well distributed over all continents, in high cultures as well as in the music of contemporary stone age cultures. Perhaps it develops wherever men play with the timbres of speech sounds. As Demetrius (first century) said: "In Egypt the priests hymn the gods by means of the seven vowels which they utter in due order; and the sound of these letters is so euphonious that men listen to it in preference to pipe and lyre. So if we do away with this clashing of vowels we simply take away the melody and music of speech" (Stanford, 1967:58).

Music for the jaw harp (mouth harp) transposes the complex play of speech sounds to an instrumental idiom. The jaw harp as we know it consists of a metal tongue in a frame. It is held against the open mouth and teeth and strummed, while the player changes the shape of his mouth cavity in such a way that groups of partials or individual partials are reinforced. Other jaw harps, especially in Asia, are made of bamboo and have a bamboo tongue, but this form has declined in favor of the imported Western jaw harp with its louder metal tongue, even in such remote areas as the New Guinea highlands, where Dan Adler found that in 1965 at least one player was unable any longer to produce a sound on a bamboo jaw harp, even though he was expert on the metal variety.

The instrument is simple but its technique requires considerable practice. It may be used in several ways: to produce tunes over the drone, in which case available pitches are limited by the partials of the metal tongue and by the resonance limits of the vocal cavities; to produce varying timbres over the drone, where the individual pitches are less important than "sounds"—animal and bird imitations, broad-band timbres in the background of an instrumental ensemble, and so on. Additional timbral nuances often heard in both styles are spectral glides, a spectral vibrato produced by rapid regular change of the mouth cavity shape, and breath sounds (noise bands). Breathing on the metal tongue makes the sound somewhat louder and enlarges the basic vocabulary of contrasts by admixtures of band-passed noise.

J. F. Rock describes a triple jaw harp called ³K'a-²kwuo-¹kwuo (see fig. 58) used to instrumentalize the courtship conversations of young persons of the Na-khi in the Li-chiang area near the gorge of the Yangtze river of southern China.

> With this instrument boys and girls arrange love pacts, rendez-vous, and suicide pacts. They will play the ³*K'a-²kwuo-¹kwuo* to the very last, before the final plunge. It is only the young couples who use them and they are never without them. The *Na-khi* Jew's harp consists of three bars which are struck with the fingers and are held between parted lips, the boy or girl will breathe the words on the bamboo rods while striking them. There are many ways of playing the ³*K'a-²kwuo-¹kwuo*, in all there are said to be 77 ways. In the olden days, the ³*K'a-²kwuo-¹kwuo* could be played in the home, mothers and daughters would use them, as well as fathers and sons. Nowadays, only the shepherd boys who wander over the lovely alpine meadows and have little to do but sit under a fir tree watching their flocks, practice the art of playing them. Girls are sent up to the mountain to fetch firewood and they also often play the ³*K'a-²kwuo-¹kwuo*. Their calls will invariably invoke a response, especially if one of the shepherd boys is the girl's lover. After a time they will arrange a meeting by means of the Jew's harp whose tones, though very soft, carry quite a distance. [1940:8]

Rock goes on to describe the seven best-known manners of playing the jaw harp, which include animal imitations (cackling crane); rhythmic and rhythmic-melodic types (driving sheep; counting sheep; dog and stag chase); and the lover's code, ²Mba-²zi-²yi-¹bi ndĕr (like the bee crossing a river).

> This is in imitation of the sound a bee makes in flying. The three bars are arranged as follows: the soprano part (bar) is placed first (upper), the bass in the center, and the medium note, lower third. The player breathes on the bars into the triangular cavity in which the tongue of the instrument rests. [P. 11]

> The three bamboo bars are held firmly between the lips, the boy or girl breathes the words on the bars while striking two at the same time. First always the two outer and then the middle and lower; this is alternately continued. The words are breathed on to the vibrating tongue in such a way that, although the words themselves are not audible, the sound vibration of each vowel and consonant is quickly recognized and interpreted. They are a sort of lovers' morse code. [Ibid.:8]

This use of the jaw harp for an instrumentalized lovers' language may have been more widespread in the past. Rock cites a paper by A. W. Young from 1908, who writes "that the Jew's harp among the tribes of Burma and Assam is a serenading instrument and is held in high estimation among them. Breathing softly on their bamboo harps, the young men address love calls to

FIGURE 58. A Na-khi boy playing the
Na-khi Jaw harp ³K'a-²kwuo-¹kwuo.

the marriageable girls; and as undesireable alliances are not infrequently the result, the missionaries have prohibited the Jew's harp in their compounds" (ibid.:12).

Of course the serenade is not in Morse or any other code. Speech, poetic speech, has been further transformed, to the point where the result is neither speech nor music but an evocative buzzing continuum into which musical and semantic meanings may be projected, a breath-mouth-music-language.

THE DIDJERIDU

The didjeridu, a buzzed-lip instrument, is one of the great drone instruments of the world. It is no more than a simple tube of wood or bamboo four to five feet long, with a bore of approximately one and a half inches or more. It may be fitted with a mouthpiece of wax, but it is playable just as it stands.

It is played by professional musicians among the aboriginal tribes of northern Australia who study the instrument from an early age, and its chief musical function is to provide a drone for the songman, another professional musician, who leads, sings, and plays rhythm sticks.

The total technique of the didjeridu is formidable, and good didjeridu players, "pullers," are much sought after by songmen, and appreciated by their audiences.

The instrument has been investigated by Trevor Jones (1956, 1957, 1965, 1967), who also took the trouble to learn how to play it. Jones has made a phonograph recording (Wattle Ethnic Series 2) which demonstrates its sounds and the most important playing styles.

Only two or three pitches are available on a didjeridu, usually the fundamental and an overblown note a tenth above. The pitch of the fundamental ranges from the D♭ immediately above cello low C to the G above, and is determined of course by the length and bore of the tube. Many of the didjeridus I have heard are pitched near low E, below the bass clef; the instrument in my possession is in E, slightly sharp.

Some styles of didjeridu playing require circular breathing; that is to say, the player snatches a breath through his nostrils while simultaneously buzzing his lips, using up a small amount of air which has been stored in his cheeks and pharynx. This technique is not unknown to wind players in the West, especially players of tuba and contrabass trombone, and it is a necessary accomplishment for playing the *launedda*s of southern Italy and Sardinia. Nevertheless, it is not easy to learn, and requires much practice.

Another very sophisticated manner of playing requires the player to hum a pitch a tenth above the fundamental while simultaneously producing the fundamental by buzzing the lips. The resultant sound is a very full-bodied major triad whose root is the fundamental tone. This technique has been used occasionally by trombonists and tubaists in Western contemporary music, but nowhere has it been more highly developed than in the didjeridu playing of certain areas of Arnhemland.

The sounds of the didjeridu are a limited system of discrete sounds, with a number of important nuances. The limited nature of the system makes it valuable to study, because timbre, timbre sequences, and timbre contrasts are the essence of it. In spite of the very different manner of tone production the music has many points in common with that of the jaw harp—spectral glides, sounds that are vowel-like and consonant-like, relationships based upon the harmonic series.

The timbre system is best expressed in terms of what the player does with his lips, tongue, cheeks, vocal chords, and breath. The main sounds then are (following Jones):

Fundamental
(*a*) basic sound
(*b*) smooth (tongue well back)
(*c*) nasal (tongue forward)

Overblown sound
(*a*) basic sound (tightened lips)
(*b*) hooted (hard tongued)
(*c*) spat (staccato, soft tongued)

Chords
(*a*) blow fundamental, hum tenth
(*b*) blow fundamental, scream
augmented eleventh above

Fake fundamental
Sounds fifth below funda-
mental (very loose lips)

In addition there is a great deal of cheek and tongue movement, one result of which is an effect of continuous spectral glide on the fundamental pitch. Some notes are hummed and gurgled to make an arpeggiated sound, with

FIGURE 59

individual pitch components entering sequentially. These glides and "arpeggios" contribute greatly to the organization of the rhythmic patterning of the drone part, although they are difficult to notate in transcriptions.

The fundamental can be raised a whole tone by hard blowing, and one often hears the beginnings of rhythmical patterns articulated in this way. Tonguing is often quite elaborate, and in some styles double and triple tonguing are common under difficult musical circumstances.

The art of the didjeridu player is in combining the sounds of this limited pitch and timbre vocabulary, and naturally some players are better than others. Weaker players may provide the singer with a basic drone with little or no rhythmic structure; better players invent elaborate rhythmic patterns, often in three against the duple of the rhythm sticks, and create extremely interesting "breaks" between verses and at other structural points in the music.

A transcription of Groote Eylandt didjeridu playing by Jones (1965) is reproduced in figure 59. The circles denote hummed pitches, the boxes, hooted notes. Check marks show snatch breaths. The quarter note of $\frac{3}{4}$ meter is equivalent to the dotted quarter of the $\frac{6}{8}$ and $\frac{9}{8}$ meters, and Jones has fitted all temporal events to those meters. The offbeat rhythm sticks are exceptional, and the melodic content of the song, while not atypical, is by no means the only melodic style to be found in Australia or even in Groote Eylandt.

Much of the music is repetitive, and some patterns may be repeated many times, marked by triangles in the transcription. An important "break" begins at the third measure, where hooted notes and fundamental with hummed tenth alternate in a complex pattern. The bore of the didjeridu used in the composition is such that the overblown note is a ninth above the fundamental. Jones relates this to the fact that Groote Eylandt didjeridus are more conical than other Arnhemland instruments.

Any large-bore instrument can be used as a megaphone or voice mask, and it seems odd that didjeridu players do not use the instrument in this manner. On the other hand the rhythm patterns are often articulated with such vigorous tongue and cheek movements, along with vocally produced hums, gurgles, and screams, that the overall sound has a speechlike aura. One would like to know whether the patterns are remembered as nonsense words, corresponding to the mnemonic drum words of India and parts of Africa, or if they are remnants of archaic and/or secret words.

The elegance of didjeridu music may not be apparent to listeners who are not attracted to drone music, but for those who are it will repay close listening, for it is one of the real summits of the art, rich in detail, highly ordered, remarkably full of sudden turns and surprises, witty changes, and subtle interplay with the singer and his percussion sticks.

DRONES AND KLANGFARBENMELODIE

As soon as one looks beyond Western music one sees a profusion of instruments designed especially to produce drones. There are double flutes with one pipe a drone; double oboes; and double clarinets. Bowed-string instruments often have one or two drone strings. The *tanpura* or *tambura* of India, a plucked-string instrument, produces only drones. Among the struck-string instruments are the *santur* (Persia) and *cymbalum* and *santuri* (Greece). The tuned drums of India and some of the drums of the Near East have an obvious pitched fundamental that often serves as a drone. The Indian drums, *tabla* and *mṛdangam,* are carefully tuned to coincide with other drone instruments of the ensemble. And among the buzzed-lip instruments are the great Tibetan trumpets, *zäns-dun* (eight feet long) and *rag-duṅ* (up to sixteen feet long), often used in pairs.

Descriptions of drone music invariably comment that it is a kind of primitive polyphony. Much of it is: a drone pitch, usually lower, is accompanied by a more or less ornate melodic line, and we are aware of two layers or strata of sound. But that description does not go far enough. The question is, are there other functions for the drone? It is interesting that drones tend to be pitched low, and some musics have gone to extraordinary length to design such drone instruments: the sixteen-foot Tibetan trumpets; an Egyptian reed pipe more than five feet long (Hickman, 1958:18, photo). The lower the fundamental pitch the easier it is to perceive individual lower partials of the spectrum. In addition to the tendency to extend toward the bass, a large number of instruments used for drones have exceptionally rich partial structures: bagpipes, trumpets, didjeridu, for example. The Indian tambura uses very thin metal strings, and the projection of the harmonics is further enhanced by placing tufts of wool or silk at certain points between the strings and the main bridge.

What are we to make of all this? More than pitch must be involved. Polyphonic layering is certainly important, and in some musics it may be the only significant effect of the drone, but in other musical styles the drone partials are in the foreground enough to form an important musical element.

Indian musicians speak of the sound of the tambura as the ambient in which the music moves. Not only is it a multiple drone of four strings, it is played in a manner that brings out the harmonic mass and obscures the individual fundamental pitches: rhythmically unpatterned and softly, with the fleshy part of the fingers. Its total sound so merges with the sound of the solo instrument that it is extremely difficult to hear any individual elements, even when watching a tambura player.

No overview of drone music can neglect this conception of the sound as ambient. One can relate it to Western concepts of chord and bandwidth as long as one also finds room for the notion of spectral change. For the characteristic progress of the music of the tambura drone is an unbroken undulation of timbre, a lazily flowing stream of sound, with subsidiary currents and cross currents curling in and out of the main channel. This attitude toward sound is important to Rush's *Hard Music* and Rabe's *Was??* and the "timelessness" of both compositions is closely tuned to the idea of a stream of unmetered sound.

The conception of drone music as an ambient or an environment is at the root of much contemporary music in this genre and is, I think, only at the beginning of a long and fruitful development. If we are interested in the "whole sound" we shall have to use that whole sound, even if we need to give up some of the frenetic darting about from pitch to pitch, so typical of our music, in order to hear and appreciate other musical dimensions. The development and elaboration of polyphony and harmonic root progression has given us opportunity to create music of unsurpassed beauty and complexity, but at the expense of restricting and constricting rhythm and timbre.

Varèse's dream of a living, unbounded music—"the corporealization of the intelligence that is in sound"—and Schoenberg's conception of a melody of timbres, are reverberations of attitudes that are deeply imbedded in Eastern music and musical thought. The *Māṇḍūkya Upaniṣad*, quoted above, says, "This identical Atman, or Self, in the realm of sounds is the syllable OM," and this statement is meant literally. Tibetan monks (Crossley-Holland, 1954) are said to find the counterparts of the sounds of their instruments in the sounds of the body: thudding (big drum); clashing (cymbals); wind blowing through a forest (conch); ringing (bells); sharp tapping (hand drum); moaning (oboe); shrill moaning (thigh-bone trumpet); bass moaning (long trumpet). A concern with numinous aspects of sound is at the center of Tibetan religious thought and "the transforming or integrating power of sound is at the very root of Tibetan religious music" (Crossley-Holland, 1954:459).

There is more than a hint of the numinous in the final sentences of Schoenberg's vision of klangfarbenmelodie, written sixty years ago:

I cannot readily admit that there is such a difference, as is usually expressed, between timbre and pitch. It is my opinion that the sound becomes noticeable through its timbre and one of its dimensions is pitch. In other words: the larger realm is the timbre, whereas the pitch is one of the smaller provinces. The pitch is nothing but timbre measured in one direction. If it is possible to make compositional structures from timbres which differ according to height (pitch), structures which we call melodies, sequences producing an effect similar to thought, then it must be also possible to create such sequences from the timbres of the other dimension from what we normally and simply call timbre. Such sequences would work with an inherent logic, equivalent to the kind of logic which is effective in the melodies based on pitch (klanghöhen). All this seems a fantasy of the future, which it probably is. Yet I am firmly convinced that it can be realized. I am convinced that [it] would dramatically increase the sensual, intellectual and soul pleasures which art is capable of rendering. I also believe that [it] would bring us closer to the realm which is mirrored for us in dreams; that, in fact, it would expand our relationships to those things which seem not yet alive to us by giving from our life to the life which appears dead to us only because we have so little connection with it. [1911:471]

5

Klangfarbenmelodie: Problems of Linear Organization

Schoenberg's argument for klangfarbenmelodie (1911), just quoted, rests upon analogy: if melodies based upon pitch have an inherent logic, then it must be possible also to create such sequences from what we normally and simply call timbre. He views pitch as an aspect of timbre: "It is my opinion that the sound becomes noticeable through its timbre and one of its dimensions is pitch. In other words: the larger realm is the timbre, whereas the pitch is one of the smaller provinces. The pitch is nothing but timbre measured in one direction." Nothing is said about the nature of relationships between timbres that might be equivalent to the inherent logic of pitched melodies, and even today, after sixty years of experience with such sequences, composers and theorists have written little or nothing about the "logic" of that rather large variety of linear formations usually tagged with the convenient term "klangfarbenmelodie."

Schoenberg's own conception of klangfarbenmelodie apparently changed little over the years, for J. Rufer quotes a Schoenberg letter of 1951 which grapples with the problem again. After remarking that "Webern's composi- tions realised only the smallest part of my conception of klangfarbenmelodie," he goes on to say that what might appear to be klangfarbenmelodie in his own works is always some kind of polyphony, and that these isolated instances are not really melodies. Melodies require a constructive unity, an organization, and, "in my conception such [klangfarbenmelodie] forms had become something new, for which as yet there is no description, because, indeed, they do not yet exist" (1969:367). The hopes of 1911 are muted, and rightly, for the analogy between klangfarbenmelodie and melody cannot extend very far. We have no intervals of timbre which correspond to pitch intervals. Pitches come in systems in Western music, and this is reflected in our scales and tuning systems. Intervals are carved out of the pitch continuum, stablized by use and tradition, and they are transposable. These intervals and their relationships, a small selection from the infinity of those available in the continuum, provide the basic contrasts necessary for melodic construction and organization. Pitch is abstracted from the whole sound, as Schoenberg points out, but what is important to melody is a system of pitch intervals.

When Schoenberg says that "pitch is nothing but timbre measured in one direction" he is correct as long as the reference is to a continuum of pitch. It is often very difficult to know whether one is hearing a slight change of pitch or a slight change of timbre (see Theories of Timbre and Pitch in Chapter 2), and even simple tones, sine tones, have a timbre that one can discriminate along a dull/bright scale. This timbral dimension of pitch is most obvious when the music shows neither intervallic pitch structure nor discrete changes, as in some of the glissandi passages in the string music of Penderecki and in certain electronic compositions.

What is important is not whether sequences of different sounds are melodies in the same sense that sequences of different pitches are melodic, but whether there is an "internal logic," meaning musical sense, to be found in such timbre sequences. Analysis is difficult because timbre is rarely to be found without pitch; indeed, there are fewer instances of any sort of klangfarbenmelodie than one might think; nevertheless, as long as one does not expect a pure pitchless klangfarbenmelodie, except in the abstract sense of a limit, and as long as the idea of linear organization does not require absolute adherence to the concepts of pitched melodies, some progress in analysis is possible.

MELODY NUANCED BY TIMBRE: A UNIVERSAL PRACTICE

Western musical practice has left many details of timbre to the performing musician. From the multitude of fingerings at his disposal a woodwind player chooses the one that fits the musical situation best; his decisions are made on the basis of pitch, loudness, ease, practicality, and propriety of timbre. The essential structure of the music is felt to be in pitch relationships as they expand into motive, phrase, section, and so on, and it is expected that the player will use nuances available on his instrument, including timbre contrast and timbre continuity, to express those felt relationships. There is nothing new here, except that it is worth mentioning that good players tune timbres in the same way that they tune pitches—to fit precisely the musical situation as they understand it.

Under such circumstances timbre is hardly an independent dimension, but our attitudes toward timbre, and our ways of using it, are a consequence of the development of Western music over a period of several hundred years. Tempo, rhythm, articulation, ranges, construction of instruments—everything has developed to favor pitch relationships, right up to the invention of electronic music, where ad hoc pitch relationships and pitch systems, fast successions of contrasting timbre, control of the spectral envelope, time envelope, and attack time have become possible.

One can imagine situations where a melody might be of primary importance but where the timbral nuance of each pitch is composed. I know of no Western music like this, but the zither music of China, as played on the *ch'in,* gives a glimpse into the possibility of detailed control of timbre on a note-to-note basis.

The ch'in is an unfretted seven-string instrument with inset studs marking the nodes of the strings. It was invented in very ancient times in China, and for centuries, even up to comparatively recent times, was favored by scholars and literary men. Timbral variety is obtained by various manners of plucking, various vibrato techniques (there are said to be twenty-four varieties of vibrato!), harmonics, double harmonics, damped and open strings, and more.

Hsi K'ang (223–262) wrote, "Now the tones tinkle like jade pendants or crystal beads, then they swell in volume, grand and abundant . . ." (van Gulick, 1969:92). "They are variegated and colourful like chequered feathers, they are like the echo left by a gentle breeeze lightly blowing" (ibid.:104).

R. H. van Gulick, who translated many of the treatises and instruction books, and who also learned to play the ch'in, explains the technique of the instrument in his *The Lore of the Chinese Lute:*

Through the delicate structure of the lute, the strings respond to the most subtle nuances in the touch. The same note obtains a different colour when it is played with the thumb or with the forefinger of the right hand, and the timbre changes according to the force with which the string is pulled. This applies especially to the technique of the left hand: beneath the nimble and sensitive fingers of the expert player the strings show a wealth of unsuspected modulations. The high notes may either have a dry, almost wooden sound, or they may be sharp and metallic, and in another passage the same note may be clear and tinkling, like a silver bell. Low notes may be broad and mellow, or so abrupt as to be nearly rattling. [1940:106]

These timbres, and a great many more, are notated by means of a tablature that tells the player what to do. The string is given by a number. A different number indicates the pitch by means of the number of the inset stud (*hui*). Another number indicates the *fên,* imaginary, small subdivisions between any two successive studs. In addition there are about two hundred *chien-tzŭ* which specify manner of excitation: which finger of the left hand, which finger of the right hand, direction in which string is pulled, and many other symbols, some calling for a sequence of complex actions; one example is the symbol *chuang:*

immediately after the string is plucked by the right hand the left moves quickly up and down the string to the right (above) of the indicated note (Kaufmann, 1967:282).

Kaufmann gives an illustration of one of the tablature symbols to show how the information is coded (fig. 60). The symbols that comprise the tablature specify pitch and timbre in a precise, and quite economical, way. In principle, the macrosymbols of the tablature are no harder to read than much of our own notation. Van Gulick tells of watching an experienced ch'in player hum the music as he looked at a new tablature manuscript.

FIGURE 60

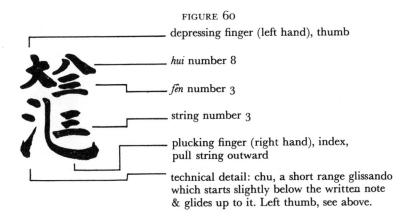

depressing finger (left hand), thumb

hui number 8

fēn number 3

string number 3

plucking finger (right hand), index, pull string outward

technical detail: chu, a short range glissando which starts slightly below the written note & glides up to it. Left thumb, see above.

Like any notation, the tablature leaves something unspecified. There are symbols for "fast," "slow," "loud," "soft," but rhythm is ordinarily not notated at all. One has to know the rhythm of the tune in advance.

One of the more famous zither tunes, *The Spell of Puuan*, has been transcribed by L. Picken (1957) into Western notation. A performance of the composition by Ja Fushi is available from the Library of Congress, Music Division, Order I095, 3 sides: 1*b*, 2*a*, 2*b*. The same performance is to be found in the UNESCO program tape, *The Music of Chinese Philosophers and Poets*, written by Picken, and distributed by Broadcasting Foundation of America. I have transposed and altered Picken's transcription slightly to conform with the music played on the UNESCO tape (fig. 61).

Reading the transcription gives one no conception of the sound of the music, only the pitch and rhythm. In fact, the transcription into Western notation does not reflect the musical sense. This drives home the fact that the succession of timbres and nuances of pitch, vibratos and beats (between simultaneous harmonics) have an importance that seems to transcend ornament. All writers on ch'in music have emphasized the importance of single sounds, and it is true that two successive sounds almost always differ in

FIGURE 61

timbre in some way. Nevertheless the essential musical structure and continuity is in the melody; the timbres, which go to the very limit of technical possibility, are all in the category of plucked strings, except for some percussive and rubbed sounds. The music uses timbre expressively, and the tune is the vehicle for a series of refined timbral gestures. They are precisely specified, and in that sense they are composed, but the chien-tzŭ are not structurally independent, and pitch is essential to the progression of the music.

When Luigi Nono requires the soloist to change his playing manner on each note in *Variante* he creates a musical situation somewhat similar to ch'in music, at least in passages where the violin is unaccompanied. In the two passages in figure 62 *n.* equals *normale,* *a.S.* equals *am Steg,* *c.l.* equals *col legno,* and *c.l. batt.* equals *col legno battuto.* A fifth manner, *flautato,* does not appear in the passages. Sequences such as these are extremely difficult, not only

FIGURE 62

Used by permission of Belwin-Mills Publishing Corp., Melville, New York.

technically—changes from *arco* to *pizzicato,* from stopped notes to harmonics, from *col legno* to normal bowing, with six dynamic levels specified—but in terms of style. Lacking a tradition for this kind of writing, it is hard for a player to grasp an overall conception of what is required. The tempi specified favor melodic continuity, but the required physical motions are close to the edge of impossibility at the specified speeds, especially in measures 119 and 270.

When Nono composes timbral and dynamic contrast with each note of a pitch succession he comes close to ch'in music, but there is an essential difference: ch'in music always has a traditional tune to structure the time sequence; Nono's melodic formations have no such preestablished structure upon which to lean. There is evidence of a plan, an intellectual scheme, and for that reason the succession of pitches and playing manners can hardly be said to be random. Missing is a perceptually meaningful organization of the sounds.

MOTIVES ARTICULATED BY TIMBRE

Nono's method in these passages and in others like them seems related to that of Webern, but for Webern timbre is always articulatory, and what is articulated is the detail of the motivic structure. In Webern's music it is difficult to find change of timbre from note to note, and easy to find motivic segments defined by a single instrumental sound. Effects of contrast between motivic units and their perceived relationship to preceding and succeeding motivic elements are very carefully calculated. An examination of his setting of the subject of the Bach *Ricercar* (see general discussion in chapter 1) is revealing. The instrumental overlap or elision on the second note of measure 6 is crucial to his conception of the internal organization of the music; therefore the transition from one instrumental sound to another at that point must be very smooth. The chart in figure 63 shows the first nine settings of the subject. The tenth and final setting is scored for unison bass clarinet, bassoon, cellos, and contrabasses. Notice how, at the elision, instruments are chosen for conjunction and disjunction of timbre and for the proper dynamic level at the particular pitch, within the available choices.

Further examination of the chart shows how Webern articulated the subject. In his letter to Scherchen about the work, he says, "Naturally, I should add that the theme should not be permitted to seem to be torn apart. My instrumentation attempts (I am meaning to speak now of the whole work) to reveal the *motivic* coherence" (Kolneder, 1961:154–155). His reading of the theme is such that the instrumentation used for the first three motivic segments is repeated for the last three segments. For example, in the first statement of the subject the sequence trombone, horn, trumpet of segments 1, 2, and 3, is repeated in segments 5, 6, and 7. The harp always doubles the

FIGURE 63

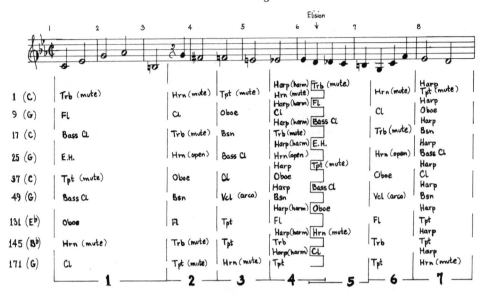

Measure	1	2	3	4	5	6	7
1 (C)	Trb (mute)	Hrn (mute)	Tpt (mute)	Hrn (mute) / Harp (harm) / [Trb (mute)]		Hrn (mute)	Harp / Tpt (mute) / Harp
9 (G)	Fl	Cl	Oboe	Cl / Harp (harm) / [FL]		Cl	Oboe / Harp
17 (C)	Bass Cl	Trb (mute)	Bsn	Trb (mute) / Harp (harm) / [Bass Cl]		Trb (mute)	Bsn / Harp
25 (G)	E.H.	Hrn (open)	Bass Cl	Hrn (open) / Harp / [E.H.]		Hrn (open)	Bass Cl / Harp
37 (C)	Tpt (mute)	Oboe	Cl	Oboe / Harp / [Tpt (mute)]		Oboe	Cl / Harp
49 (G)	Bass Cl	Bsn	Vcl (arco)	Bsn / Harp / [Bass Cl]		Vcl (arco)	Bsn / Harp
131 (E♭)	Oboe	Fl	Tpt	Fl / Harp (harm) / [Oboe]		Fl	Tpt / Harp
145 (B♭)	Hrn (mute)	Trb (mute)	Tpt	Trb / Harp (harm) / [Hrn (mute)]		Trb	Tpt / Harp
171 (G)	Cl	Tpt (mute)	Hrn (mute)	Tpt / Harp (harm) / [Cl]		Tpt	Hrn (mute)

instrument used in the seventh segment, and it always adds a unison accent at the beginning of the fourth segment. The strictness of this patterning is upset only once, in the statement beginning at measure 49, where cello and bassoon are interchanged. The interchange clarifies the relationship between subject and other voices at measure 56.

There are other niceties in Webern's instrumentation which may be gleaned from the chart. There is progress in timbre contrast from the all-brass opening statement to all woodwind to mixed brass and woodwind entries. I place the extremity of contrast in the statement at 131 (oboe, flute, trumpet), with the flute at the accent at the beginning of segment 4. The flute is in a range where it certainly has enough dynamic power for that accent, and the oboe, in that particular range, can sound very flutelike. Nevertheless, it seems slightly out of character for Webern to dispose the instruments in this way. If one looks ahead to measures 137 and 139 (fig. 64) one can hear what he may have had in mind, for between the last three notes of 137 and the first three of 139 there is a kind of musical pun. If the instrumentation had changed at measure 139 the sense would have been lost.

In episodes and counterpoints the segments are longer and contrasts more subdued. A single instrumental sound—usually strings, and note how seldom the strings participate in the instrumentation of the subject—may continue for many measures. Within any instrumental category, however, Webern continues the process of articulation. For example, the first and second violins often break a line into motives or phrase segments, and the articulation is then in terms of spatial position. He also makes much use of solo vs. section and *pizzicato/arco* contrasts.

FIGURE 64

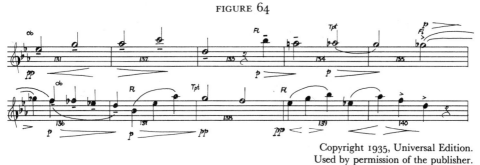

An examination of the composition as a whole reveals a larger plan for the disposition of changes of timbre. At the opening, and again at the beginning of the second half, the segments are shorter, and as the climaxes are approached instruments play for longer and longer periods of time. There is, therefore, an overall movement or rhythm of segmentation, produced mainly by means of timbre.

The fourth piece of Webern's *Five Pieces for Chamber Orchestra*, Opus 10 (fig. 65), also employs timbre for articulation, segment definition, and connection. The change of instrumental sound between harp harmonic and celeste in measure 5 is very subtle, hardly a disjunction of timbre, more a sensitive nuance. Harp and celeste also blend and blur the natural sharpness of the mandolin at its entrance in measure 5.

The most striking passage for timbre is immediately after the opening mandolin statement, for there is complication and interaction between the linear and vertical here which obscures the line while connecting its elements. I hear the second phrase segment as beginning with the viola harmonic in measure one, but, as soon as the clarinet enters on A, *pianississimo*, I tend to feel *it* as the first phrase note, and the viola B♭ as an element of accompaniment. I have a similar response when the trumpet enters *pianissimo*, *dolce*. The melodic structures that might resolve the haziness of the situation are not very strong. The C, D, A♭ that opens the mandolin phrase has a counterpart in the A, B, F of mandolin and trumpet in measure 2. The same intervals are to be found in retrograde form in the violin phrase that concludes the piece. The intervals of the minor ninth at the very end of the violin phrase corresponds to the ninth, B♭-A between viola harmonic and clarinet in measures 1 and 2. These are weak relationships, all the more tenuous when they are between different instruments, but I believe Webern is playing upon such ambiguities here: a listener may follow either the tone color or the pitch, or, as the composition becomes familiar, both, as elements of a perceived whole.

FIGURE 65

IV.

FIGURE 66

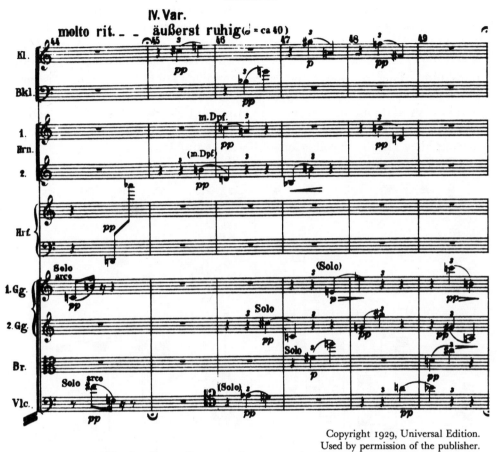

Webern's middle-period works show more systematic use of timbre as a structural element. Short melodic motives of from two to four pitches are delineated by separate instruments or by manner of playing, and the timbre placements often reflect the skewed symmetries of his pitch organizations. If the motivic organization is in two-note groups, as in Variation IV from the second movement of the *Symphonie*, Opus 21 (fig. 66), then there is a remarkable play of fast-changing instrumental sound. The motivic structure is obvious here, and the larger organization relies upon change of instrumental color to emphasize some pitches more than others. This quality of the music is most noticeable when a pitch is repeated after a rest or interpolated note. For example, notice how the G♯ of the clarinet in measure 47 is repeated by the violin in measure 48, and how the viola F at the end of measure 47 is repeated by the clarinet in measure 48. We hear the repetition, the melodic/motivic situation in which it is disposed, and we hear the difference in timbre. If such note repetitions with change of instrumental color are highly patterned then the timbre itself becomes integral to the formal organization. A graph of Variation IV, made by Ken Timm (fig. 67), shows Webern's overall patterning. If one reads, say, the A below middle C from left to right, it is easy to see how a pitch is repeated in different timbres, and how the plan works. I do not wish to maintain that the only important changes of timbre are across repeated notes, but it must be clear that repeated pitches have a special status, because they are the simplest sort of linear channel.

FIGURE 67

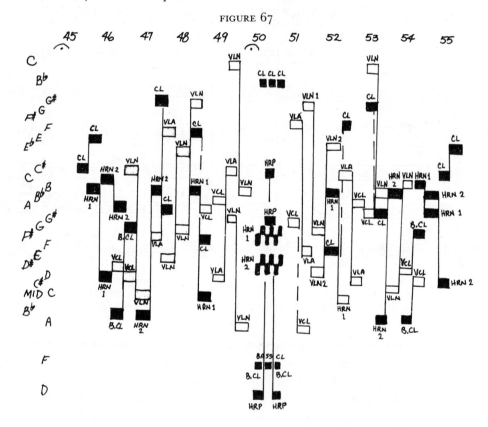

CHANNELS, MOTIVES, MELODIES, TIMBRES

There is nothing recondite about the concept of channel in music, and an illustration can quickly show how it works. Figure 68 is a transcription of music sung and played by a member of the Ba-Benzelé from the African Congo. The timbre of alternate notes is either voice or the sound of a small whistle. The music is a totality, and the effect is remarkably continuous, but one can also listen either to the whistle line, with its equal spacings and different timbre, or to the vocal line. Persons trained in European musical styles are likely to hear the composition as polyphony—a tune below an ostinato drone. Channeling is reinforced because the whistle pitch is always repeated, and because, except for the second note of each group (occasionally the first), the voice has a different *tessitura* of a few step-related pitches. The musical interest comes largely from the tension between the phrase heard as a whole and the phrase heard as a composite of two channels. When the musical texture is hocketed, as in much of Webern's music and in this Ba-Benzelé tune, the ambiguity about linear connection which is a consequence of the numerous rests can be turned to the composer's advantage: timbre can be employed as a structural element to create effects of melodic continuity or disjunction, ambiguity and its resolution.

FIGURE 68

Kenneth Gaburo has worked directly with channel ideas in some of his solo instrumental compositions, for example, *Inside*, quartet for one double-bass player (1969) (fig. 69). The quartet idea is characterized by: (1) voiced phonemes derived from the word "inside"; (2) extravocal sounds, for example, kissing, rolled *r*; (3) bass effects that employ the use of the strings; and (4) bass effects that do not employ the use of the strings. Timbre and dynamics are carefully specified, often by means of special symbols. Pitch is specified more broadly. Beyond the four-part plan, other channels form and dissolve: all /s/ sounds; all voiced phonemes; all bowed sounds—any sufficiently similar set of sounds in the context of the particular phrase. The "parts" are neither continuous nor in fixed ranges, and the overall effect is of a klangfarbenmelodie of pitched, unpitched, vocalized, and mixed sounds which often transforms itself into polyphony. The phonemic elements of the word "inside" provide a more than skeletal structure—they are an important component of the thematic material of the composition, a set of fixed points in the timbral continuum, around which the other sounds are organized and shaped.

Channeling of pitch was first studied as a psychoacoustic phenomenon by G. A. Miller and G. A. Heise (1950). They alternated two oscillator frequencies five times per second and found that when the frequency difference was large they sounded like two sets of unrelated, interrupted repeating tones. The authors defined the trill threshold as the transition point between the hearing of a trill and the hearing of two separate pitches. This

FIGURE 69

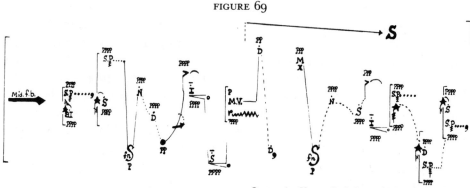

© 1974 by Kenneth Gaburo. Used by permission.

transition region is of special musical interest, and the trill threshold idea is a good deal broader in musical application than musical trills. "Channel" is a better term for these processes, in spite of the fact that it has been applied with abandon in a great many disciplines, to mean a great many things, ever since the first papers on information theory. In experimental psychology it has been used in the sense of discrete sensory channels, such as left ear, right ear. A. S. Bregman and J. Campbell, discussing some experiments on the discrimination of temporal order, have given a definition of channel for psychoacoustic experiments which fits the broader musical situations: ". . . a channel may be defined as a subjective auditory stream which is segregated perceptually from other co-occurring auditory streams. We assume that attention cannot be paid to more than one such channel at a time, i.e., that the *apparently* simultaneous streams produced by channelization have the same properties as *actually* simultaneous streams sent to separate ears" (1970:3).

A later paper by Bregman (1971) "On the Perception of Sequences of Tones" substitutes "stream" for "channel." He finds that the tendency for high and low tones to split apart starts at a speed of one per second, that glissandi to successive tones inhibit the splitting even when the glissando goes only part of the distance to the next tone, and that (as we might expect) pitch ranges and tempi are important dimensions.

Channel concepts are probably crucial to any theory of klangfarbenmelodie and can be useful even at our present stage of development. Experiments such as those reported by Bregman ought eventually to bring us closer to language for a theoretical statement about the kinds of timbre successions we are creating. It may well turn out that any klangfarbenmelodie is in essence compound, with a hidden, and often not so hidden, polyphony close to the foreground of our attention. It seems certain that a klangfarbenmelodie having any musical interest cannot be a mere linear succession of contrasting timbres. There must be internal organization on several levels, and channeling may be significant on one or more of those levels.

The hocketed polyphony that Webern exploited in the works of his middle

FIGURE 70

FIGURE 71

years has a traditional base, the use of a single instrumental sound for each
musical motive, and radical results on many levels, not the least of which is
the channeling that is a consequence of his pitch and timbre patterning. The
Symphonie, Opus 21, has timbre strongly bound to motive, and this is largely
true to of the *Concerto for Nine Instruments*, Opus 24, although there are a few
passages where the timbre changes from note to note, for example in measures
41–44 of the final movement (fig. 70).

The notes F, E, G♯ of measures 41–42 stay in a single channel for
motivic/melodic reasons, although the first two are for muted horn and the
last for trombone. The change to trombone articulates the motive—its attack
quality, loudness relationships among its parts, and so on—and the trombone
notes of 42 and 44 hardly form a channel. Had the trombone pitch in 44
repeated that of 42 we would be more aware of an identity, and a trombone
channel would have been more likely to form.

Repeated notes of different timbre are important to Webern in the *Variations
for Orchestra*, Opus 30, and this may bear some relationship to the fact that in
this work the surface of the music—that which we follow—is not as directly
related to motivic units as in the *Symphonie* or the *Concerto*. Consider measures
9–11 of the *Variations* (fig. 71). The trombone F at the end of measure 10 is

FIGURE 72

repeated by the cello at the beginning of measure 11. The two F's tend to form
a channel different from the individual melodic fragments of the trombone
and cello parts. This "confusion" and interlocking is an essential element of
the *Variations*, and it appears on almost every page of the score. In figure 71
there is the same kind of timbral play on low G♯/A♭ between contrabass,
cello, and tuba in measures 9 and 10.

This timbre interlocking, and it is always more than the repetition of motive
in a different timbre, may be heard from the start of the composition,
measures 3–7 (fig. 72), the first polyphony of the piece. In measure 3 the viola
B-D repeats the B-D of the oboe, and, with less complication, the bass clarinet
in measure 7 repeats the G of the cello and goes on with a retrograde form of
the cello pitches.

While such details are of more than passing interest they are part of a
compositional procedure that can be seen to better advantage in longer
passages, such as between measures 74–99 (fig. 73). I have transcribed the
passage so that pitches repeated in a different instrumental color are written
in larger notes; numbers from 1 to 12 are of the various row forms employed.

In terms of row form, the first two measures have a melodic fragment for
flute whose inversion is given by cello and clarinet. The row forms yield the
same pitch, A, at the end of the second measure, so there are two separate
melodic channels. The two measures are, however, also a single melodic
entity, a klangfarbenmelodie: flute to cello to flute to clarinet to flute. The
repeated A is a cadence of this klangfarbenmelodie which is also a pitch
melody, and moreover a compound pitch melody and a compound klangfarbenmelodie, because it is articulated (timbres, rests, ranges, motives) on all of
those levels. Repeated listening and a study of the score will show how
elegantly Webern has built timbre into this music, and how he has created a
klangfarbenmelodie with a large vocabulary of sounds while remaining true to
the ideas about motivic organization, somewhat extended, which he had
developed in the *Symphonie* and other works of his middle years.

FIGURE 73

WHEN THERE IS NO MOTIVE

When motivic organization is placed in the background there is naturally less opportunity for Webern's kind of interplay of tension and relationship between melodic and timbral channels. If the music has rhythmically profiled phrases, and if there are perceived goals of melodic motion, then the total effect of timbre contrast is likely to tend toward nuance, as in ch'in music. If the linear-melodic elements are weakened then attention moves to individual sounds, as in much of the music of John Cage.

Nono's music at number 7 of *Il Canto Sospeso* (p. 62) (fig. 74) is almost a textbook instance of such a klangfarbenmelodie. There is no overlapping of instrumental sounds here; there are no rests, and at the rather slow tempo, little tendency to split into channels, although the pitch range is rather wide.

Morton Subotnick has experimented extensively with electronically generated klangfarbenmelodie in such works as *Touch*, *Silver Apples of the Moon*, and more recently (with an improved synthesizer built by Don Buchla) in *Sidewinder*. During the final minute or two of each record side of *Sidewinder* there are short passages of klangfarbenmelodie, some of which are rhythmically profiled, with a more or less regular pulse; other passages are more fluid,

FIGURE 74

Used by permission of Belwin-Mills Publishing Corp., Melville, New York.

with little or no suggestion of underlying regularity of pulse. The faster, rhythmically profiled passages allow for channeling much more than the slower, rhythmically irregular and less-profiled passages.

Most of the sounds involved are rather short, which in itself makes for limitation. (Notice how Nono, in figure 74, has used the effects of short and long notes.) Nevertheless, Subotnick's electronic klangfarbenmelodie are successful, I think, because the repertory of attack envelopes and spectra of the various sounds is large enough to generate convincing contrasts.

Any ensemble composition by John Cage, but especially those involving an ensemble of instruments, is likely to present occurrences of linear timbre contrast with a guaranteed absence of motive. Cage's devotion to indeterminacy has been well publicized, but an examination of his ensemble compositions shows that in many of them the composer exerts considerable control over the progress of the music. *Atlas Eclipticalis* (1961–1962) has a time window of about five seconds. Players enter, drop out, count rests, and play in certain prescribed ways.

To be sure, as in his other ensemble works, Cage stipulates that the eighty-six instrumental parts may "be played in whole or in part, any duration, in any ensemble, chamber or orchestral, of the above performers; with or without *Winter Music*" (1962:30). Nevertheless, the parts are written in such a manner that a group of orchestral size could perform in a unified way (see fig. 75).

Each part is written in space equal to a time at least twice as slow as clock time. Arrows indicate 0″, 15″, 30″ and 45″. Space vertically equals frequency. Since equal space is given each chromatic tone, notes not having conventional accidentals are microtones. Specific directives and freedoms are given regarding duration of tones. Loudness is relative to the size of the notes. Tone production is never extraordinary. Percussion parts are a graph of the distribution in space of the instruments, as various and numerous as possible, chosen by the performer. [Ibid.]

In addition to these constraints the instrumental parts for violins and French horns are divided into high range and low range. That the work has been "scored" may be inferred from an examination of pages from two individual instrumental parts, illustrated in figure 75. In spite of these constraints the moment-to-moment succession of instrumental sounds is thoroughly uncontrolled. One can listen to the sounds themselves and be alert for miracles to happen. A particular performance may thus be used as laboratory or fieldwork for studying timbre divorced from motivic/melodic organization, and that is the sense in which I have used the recorded performance by the Ensemble Musica Negativa, directed by Rainer Riehn (Deutsche Grammophon Gesellschaft 137 009).

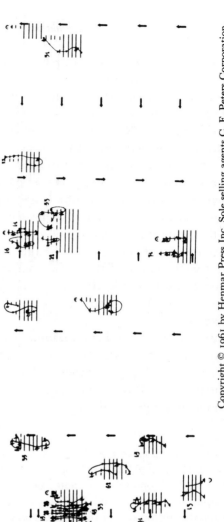

FIGURE 75

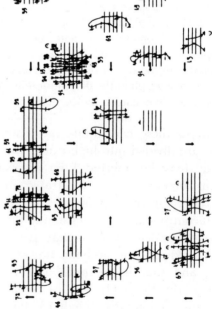

This performance combines Cage's *Atlas Eclipticalis* and *Winter Music* with *Cartridge Music* electronics, as suggested by the composer, and uses fifteen solo instruments, percussion, and four pianists. There are several fine moments of klangfarbenmelodie, one of which, a ten-second sequence starting 2:01 after the first sound, is transcribed in figure 76. The arrows above and below some of the notes indicate microtonal sharpening and flattening, as prescribed by the composer. The numbers from 1 to 9 show time in seconds. All short notes are transcribed as grace notes, and long notes are given approximate durations in seconds.

FIGURE 76

A glance will show that, either by chance or rehearsal, this passage has considerable pitch structuring: the repeated E's; the channeling into a high line and a low line; the pitch contour of the whole phrase. As a klangfarben-melodie it is more complex than the example from *Il Canto Sospeso* because of the overhang of the longer notes. This quality of shingling, which was discussed in chapter 2 as characteristic of much of Varèse's music, always tends to reinforce vertical relationships, and in this case to cause the music to hover on the edge of polyphony. It is precisely the ambiguity between melodic, harmonic, and polyphonic tendencies which raises the idea of klangfarben-melodie above mere formula. This ambiguity, and its complex musical effects, is nicely projected by the Musica Negativa players in their shaping of this beautiful and expressive phrase.

The same ambiguity between linear, vertical, and polyphonic elements is in Stockhausen's *Kontra-Punkte* (1952–1953), this time with more precise specification by the composer. Although *Kontra-Punkte* is often cited as a work of extreme pointillism there are many passages, more and more as the work proceeds, where a single instrument continues for many notes. In a radio interview with Eric Salzman (1964) Stockhausen commented on an essential problem of nonmotivic klangfarbenmelodie: "The whole thing is that if you make everything different everything becomes the same. We know this from the work of Boulez, for example, the *Structures* for two pianos, or from other works of that time of the Frenchman Fano, or the Belgian Goywaerts; so I tried to avoid this, and in *Kontra-Punkte* I think there are differentiations inside of a pointillistic form."

Some of these differentiations come about through what Stockhausen in the same interview characterized as composing equalities between sounds: if a sound is very loud it could be made shorter; if very soft it could be equalized by being made longer. Projecting timbres could be controlled by registral and other means. Being equal in this sense would not mean being undifferentiated.

In the first twelve measures of *Kontra-Punkte*, given in-score in figure 77, one notices that more traditional means of balancing the ensemble are also used, for example, the writing of the dynamics for trumpet and clarinet in measure 6. There are also some very interesting changes of timbre on a single pitch, for example, the B♭ below middle C: cello, in measure 5; to bassoon, measure 7; clarinet, measure 11; piano, measure 12. There is more shingling, and it extends over longer time spans, than in the Cage example, and from measure 7 the polyphony is overt.

A nice touch is in measures 4 and 5, where the C below middle C is doubled by bassoon and trombone. The two timbres are matched by controlling the dynamic level of each, and the mixed timbre of trombone/bassoon is marvelously clarified by the piano at the beginning of measure 6. Similarly, the short, double-plucked G♯/A♭ of the harp in measure 5 is remarkably distinct, though *pianissimo*, like an attack separated from its continuation, the violin A♭ of measure 7. And one can observe how the chord at the beginning of measure 7 is motivated—each of its pitch elements has appeared earlier in another instrument.

The term "klangfarbenmelodie" may not appear to fit music such as *Atlas Eclipticalis* and *Kontra-Punkte*, with their shingling and channeling; on the other hand it seems that compound melodic formations and implied polyphony are the rule rather than the exception in klangfarbenmelodie. Where linear and vertical descriptions are involved there has always been considerable ambiguity, even in, say, the solo sonatas of Bach. The important thing is to understand that the use of the two dimensions in description and analysis cannot change the fact that the flow of the music is, in a most important sense, one whole.

WHEN PITCH IS NEGLIGIBLE, VAGUE, OR CLOUDED

Whenever timbre is changed without change of pitch the result must necessarily be some sort of drone, and interest can be focused upon discrete and/or continuous changes of timbre. The most famous passage of this sort is probably the crescendo on B from Berg's *Wozzeck*, where the main interest is in the swell of the loudness curve and in the increasing richness of the single pitch band. Gunther Schuller has written a short klangfarbenmelodie on the tone A 440 in the slow movement of his *Woodwind Quintet* (1958), and one of the etudes from Elliott Carter's *Eight Etudes and a Fantasy for*

Woodwind Quintet (1950) is entirely on G (fig. 78). Timbres are overlapped in a variety of ways, and through a large vocabulary of attack types Carter creates some interesting ensemble timbres.

Varèse was fond of the idea too, and passages are sprinkled throughout his works. The measures from *Déserts* in figure 79 are typical of many that capitalize upon continuous change of timbre in a brass-instrument crescendo. Here there is also the interesting overhang of the suddenly quiet French horn.

Some instruments have the potential for solo performance of klangfarbenmelodie on a single note, either through alternate fingering, change of embouchure, or both. The winner of any contest would have to be the bassoon, with its thousands of alternate fingerings and shadings, but most instruments have possibilities that have remained dormant largely because composers have not composed those timbral nuances.

Musical situations where there are no clear pitches give the clearest examples of klangfarbenmelodie. Not that pitch can be eliminated—there is always higher and lower—but in situations where the pitches are unsystematized pitch intervals are correspondingly less important to the structure. Under these "weak pitch" circumstances any klangfarbenmelodie is closer to the idea of a succession of timbral objects, a melody of "sounds." In spite of statements by Varèse and remarks by John Cage—"let the sounds be themselves"—such passages are not very numerous, even in the works of Cage and Varèse. Any sequence of "sounds" can form itself (or be formed) into an organized line if dynamics, tempo, and context permit. Natural sounds and machines sometimes present themselves in this way, and certainly one of the pleasures of being alive is to perceive and appreciate the images of our auditory landscape.

Thus, in a particular music pitch may or may not be significant. Joji Yuasu used bands of filtered white noise to produce "no-pitch" klangfarbenmelodie in his *Icon*, for five-channel magnetic tape (fig. 80). Individual sounds are defined by bandwidth, range, and amplitude, and are further differentiated by direction and movement among five speaker channels. The composer does not attempt to control attacks of the succession of noise bands, and there is no variation in rustle time.

Unpitched percussion sounds are another source of klangfarbenmelodie. In a passage from *Tempi Concertati* by Luciano Berio for cowbells, wood blocks, bongos, and tom-toms (fig. 81), each instrument group has pitch gradations, but there is no pitch system, and the melodic events, while weighted in favor of the cowbells, depend upon discrete contrasts between instrument groups.

Włodzimierz Kotoński uses a somewhat larger set of unpitched timbres in his *Trio for Flute, Guitar, and Percussion* (1962). The layout of the score (fig. 82) enhances the klangfarbenmelodie quality of the music, and, while a single instrument often plays several notes, the tendency is toward strong sequential contrast.

FIGURE 77

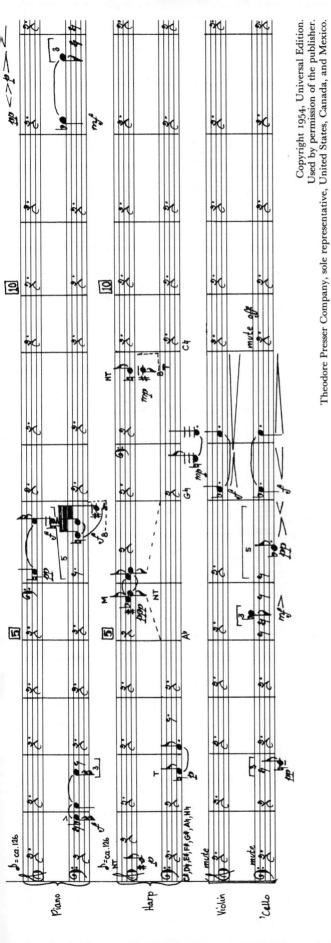

FIGURE 78

FIGURE 79

Jerome Rosen's *Elegy for Solo Percussion* has an interesting passage for log drum incorporating continuous timbre change. The player rolls along the log drum and to the rim of a snare drum. This effective phrase begins with sticks and ends with fingers (fig. 83).

A sort of Klangfarbenmelodie within a single instrumental type—gongs and cymbals—is developed at length in the central portion of Luigi Nono's *Coro di Didone* (1958), for chorus and percussion. The section calls for eight hanging cymbals of various sizes and four tamtams, managed by six players. Beaters

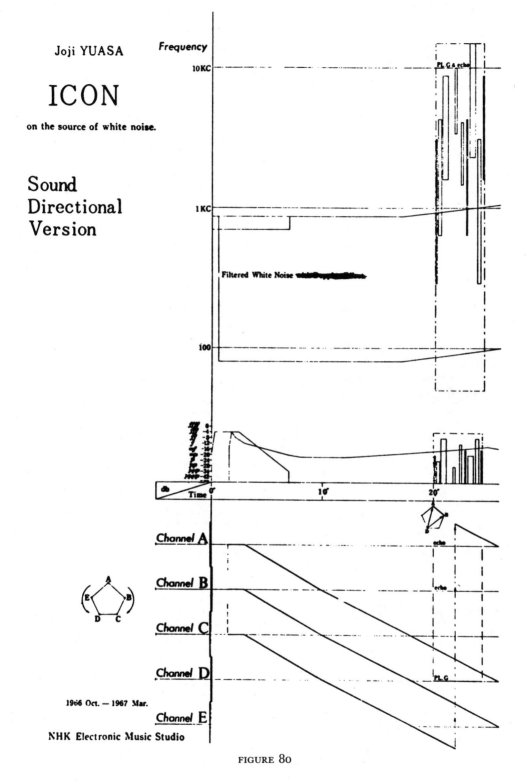

FIGURE 80

FIGURE 81

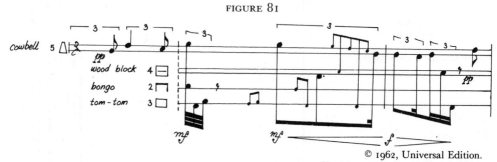

FIGURE 82

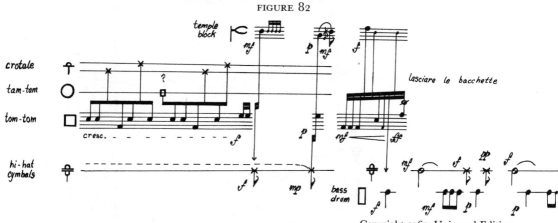

FIGURE 83

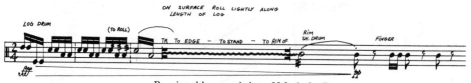

are specified and damping is precisely notated. Sometimes the ensemble is used to produce homogeneous textures, but often the instruments perform more melodically, as in figure 84.

The recent interest in the voice as a producer of a large gamut of timbres is certainly related to klangfarbenmelodie, and there is a growing list of vocal compositions where voice or voices are used specifically as sound sources for unpitched klangfarbenmelodie.

The genre reaches well back into the twenties, a period of feverish experimentation in European poetry, much of it associated with Dadaist groups. In 1921 Kurt Schwitters heard Raoul Hausmann's sound poem, *fmsbw*. He became interested in Hausmann's ideas and often recited his poem at lectures as "Portrait of Raoul Hausmann." During the twenties and thirties Schwitters composed a *Sonata in Primeval Sounds*, where the material of Hausmann's poem was transformed, developed, and eventually extended to Mahlerian length, *fmsbw* becoming part of one of the themes of this adventurous work. It is notated as: "Fümms bö wö tää zää Uu" and so on (fig. 85).

The sonata has four movements, and the complete work is almost forty minutes long. According to Ernst Schwitters, the artist's son, the work was recited from memory, and Schwitters worked it over from performance to performance, constantly changing, improving, and polishing it. Various states of the sonata were published in Schwitters's magazine, *Merz*, and the page reproduced is from the final version published in *Merz*, number 24, in 1932. A recording of the complete composition, in a performance by Ernst Schwitters, was made in 1962.

Not much can be said for the artistry of the thematic development. It is mostly bad, the kind of plodding thematic/motivic workout that had served the dullest composers of nineteenth-century sonatas. But the themes are another matter—they are full of invention, striking juxtapositions, and internal correspondences; they are klangfarbenmelodie in the modern sense of the word.

A fine example of a modern musical treatment of the voice is *Sound Patterns* (1964) by Pauline Oliveros, which uses a mixed chorus to produce a large variety of unpitched timbres. Much of the work is polyphonic, but the opening distributes a klangfarbenmelodie among the various vocal sections. In figure 86 notice the timbral glides in the first and second measures and the mixed vowel between the alto and tenor parts at the beginning of the second measure. The timbre notation here is in phonetic syllables, based upon *Webster's Collegiate Dictionary*, along with a few special markings, and proves to be extremely practical and precise for its purpose.

No doubt a background for all this expansion of vocal variety can be found

FIGURE 84

Used by permission of Belwin-Mills Publishing Corp., Melville, New York.

NB: ♪ ⃰ : *nicht über den Notenwert klingen lassen*

FIGURE 85

SONATE IN URLAUTEN"

(Sonata in Primeval Sounds — Sonate Présyllabique)

einleitung:

Fümms bö wö tää zää Uu,
       pögiff,
           kwii Ee.

1

Oooooooooooooooooooooooooooooooo,

6

  dll rrrrrr beeeee bö,   **(A)** **5**
  dll rrrrrr beeeee bö fümms bö,
    rrrrrr beeeee bö fümms bö wö,
       beeeee bö fümms bö wö tää,
          bö fümms bö wö tää zää,
            fümms bö wö tää zää Uu:

erster teil:

thema 1:
Fümms bö wö tää zää Uu,
       pögiff,
           kwii Ee.

1

thema 2:
Dedesnn nn rrrrrr,
      Ii Ee,
         mpiff tillff too,
             tillll,
                Jüü Kaa?
                (gesungen)

2

thema 3:
Rinnzekete bee bee nnz krr müü?
               ziiuu ennze, ziiuu rinnzkrrmüü,

3

   rakete bee bee.

3a

thema 4:
Rrummpff tillff toooo?

4

---

in some of Schoenberg's *sprechstimme* ideas. But *sprechstimme* itself was new only in the sense that it was notated. The sounds themselves, the manners of delivery, had been a part of stage diction and operatic style for a very long time. Indeed, one cannot find the beginnings for the modern use of the voice as a sound source; they fade into the shadows of prehistory.

FIGURE 86

PAULINE OLIVEROS

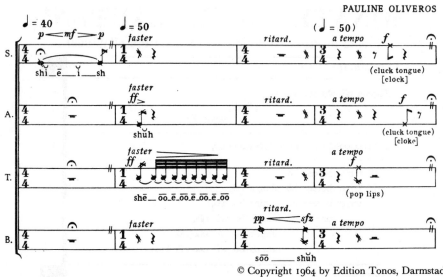

Some cultures have preserved songs whose "words" are successions of meaningless syllables. C. M. Bowra has analyzed several of these in his *Primitive Song*: "In any case senseless sounds seem to be the most primitive kind of song. They anticipate later developments by making the human voice conform to a tune in a regular way, but the first step to poetry came when their place was taken by real words" (1962:61). Bowra seems to be saying that the direction of development is always toward words and away from senseless sounds, but he offers no evidence to support that speculation, and it may well be that some senseless sounds are survivals whose phonetic elements have been worn smooth into musical pattern long after ancient meanings have been lost.

The highest development of musical but meaningless speech sounds is in the drum syllables that can be found wherever we find flourishing drum musics, especially in India and Africa. Speech syllables are used to symbolize the various drum strokes of the particular system, and in general the sound of the syllable matches the sound of the drum stroke. In highly developed musics, with a long oral tradition, such as those of north and south India, one finds, however, that the drum language tends to go its own way—in south India, for example, where there are more syllables than there are drum strokes. Robert Brown's study of South Indian drumming makes this quite clear:

The drum syllables lead a life of their own, so to speak, and the relationship with the actual drum sound, although often explicit, can sometimes be quite poetic. The stringing together of longer patterns depends on aesthetic and euphonic conditions.

"It is not beautiful" would be the immediate reaction to an amateurish linking of DIKUTAKA KIṬATAKA, and one must admit that by any criterion DIKUTAKA TARIGIḌU is better.

Therefore, the rumbling, humming, and ticking up and down of the spoken syllables has a cadence, a continuity, and a logic of expression that finds a counterpart in the logic of expression, continuity, and cadence of the drum sounds, but the up and down ticking, humming, and rumbling patterns may not be the same [1965:135–136].

One may therefore view a given sequence of drum syllables, such as the mṛdangam lesson below, as a klangfarbenmelodie of speech sounds (ibid.:15).

Lesson 14 (PI)

TA . KIṬA KIṬATAKA TARIDANA KIḌAJOṆU DIKUTAKA TARIGIḌU

    2 1   2 1   4   4 1 4 N   2 1 2   1   A   2 1   4
ā   ā   ā   ā   ā   ā

DI . KIṬA KIṬATAKA TARIDANA KIḌAJOṆU DIKUTAKA TARIGIḌU

X   2 1   2 1   4   4 1 4 N   2 1 2   1   A   2 1   4
             ā   ā   ā   ā   ā

TOM . KIṬA KIṬATAKA TARIDANA KIḌAJOṆU DIKUTAKA TARIGIḌU

O   2 1   2 1   4   4 1 4 N   2 1 2   1   A   2 1   4
ō   ā   ā   ā   ā   ā

NAM . KIṬA KIṬATAKA TARIDANA KIḌAJOṆU DIKUTAKA TARIGIḌU

N   2 1   2 1   4   4 1 4 N   2 1 2   1   A   2 1   4
             ā   ā   ā   ā   ā

Each syllable is the same length. A dot is the length of one syllable. Drum strokes are coded beneath the syllables. Left-hand strokes are beneath the line, right-hand strokes above. The syllables from the beginning of each line, /TA/, /DI/, /TOM/, /NAM/, are fundamental mṛdangam sounds under study, and they have been prominently articulated. Allowing for the fact that a primary function of this music is rhythmical, it is still clear that this lesson may be heard as a klangfarbenmelodie, that the syllables are independent klangfarbenmelodie parallel to the drum klangfarbenmelodie, and that the syllables are a remarkably precise notation of that klangfarbenmelodie.

Note that articulation is by timbre rather than by dynamic accent, and that the syllables are grouped into "words" that express that articulation. This is one of the more interesting and difficult aspects of any klangfarbenmelodie— what belongs to what—and anyone interested in fast timbre changes and their grouping into segment, phrase, and phrase group could profitably study drum languages and drum music such as this.

The lesson notated above would usually be practiced at three or four or more degrees of speed, with important effects upon articulation, segmentation, contrast, and continuity—as discussed in chapter 3. Brown explains that a phenomenon of this practice of playing everything at several degrees of speed is that

the patterns change character in a very striking way at fast speed. This is difficult to describe, but not to hear. In lesson 14, for instance [see above], the pattern at slow speed sounds like a string of six groups of four beats, which it is, with an interesting texture of damped strokes of different types, one NAM, and one *araicāppu*: [NAM = N; *araicāppu* = A]

TA . KITA KIṬATAKA TARIDANA KIḌAJOṆU DIKUTAKA TARIGIḌU

The textural accents that emerge from the juxtaposition of these timbres at high speed have a syncopated quality that can roughly be suggested:
This seems to be due to a slight swallowing up of some left hand strokes in the overall sound (which, nevertheless, is always clearly articulated), as well as the prominence of the two strokes with tonal constituents. [1965:137–138]

There is more to be learned about klangfarbenmelodie from drumming and drum syllables. I began this chapter by saying that we have no intervals of timbre, but this music has. The drum sounds in the lessons analyzed by Brown in his dissertation comprise twelve strokes. The drum syllables number twenty-nine, but, after allowance for linguistic assimilation and euphonic requirements, there remains a core of about a dozen. The drum strokes are fixed, or relatively fixed, timbres at any given speed, and their relationships are limited to those possible within a small vocabulary. (This limitation of vocabulary may be heard in didjeridu music too.) If we are to have intervals of timbre it appears that certain sounds must be carved out of the continuum, rationalized, named, learned, exactly parallel to the way in which our tradition has rationalized certain pitches and their relationships. In the West we have learned how to make musics from selected pitches. We have hardly begun to think of timbres in fixed relationships, and perhaps such limited sets of fixed timbres are not a fruitful direction for us, but those musicians who see that direction as a possibility may find that Indian music can give them more than an inkling of what such a music might be like.

6

Timbre in Texture

Texture always denotes some overall quality, the feel of surfaces, the weave of fabrics, the look of things. Words from visual and tactile sense modalities are often appropriated for descriptions of sounds and their combination: sharp, rough, dull, smooth, biting, bright, brilliant, brittle, coarse, thick, thin, dry, diaphanous, airy, finespun, flaccid, fluid, gauzy, glittery, grainy, harsh, hazy, heavy, icy, inchoate, jagged, limpid, liquescent, lush, mild, murky, pliant, relaxed, rippling, to name a few. During the past half-century more music has been written in which texture itself is primary. Some works, Krzysztof Penderecki's *Threnody to the Victims of Hiroshima for 52 Strings* and György Ligeti's *Atmospheres*, for example, are overwhelmingly texture, and, in works such as Terry Riley's *In C*, contrapuntal and tonal construction is made to yield a predominantly textural experience.

Masses, clouds, and other composite textural effects are by no means a recent invention. Debussy, Strauss, Wagner, Berlioz, and their contemporaries developed and vigorously exploited the potential of orchestral textures. But in most of the compositions of Varèse, and in certain compositions of Schoenberg, Berg, Webern, and Stravinsky, texture is central, in no way a by-product of contrapuntal, harmonic, melodic, or rhythmic processes.

In any texture some elements stand out and others recede. Completely featureless textures are rare. Louder, higher, contrasting elements attract attention, and we hear them in the foreground. The spatial terminology is quite appropriate—music without foreground and background elements is difficult to find—but there may be more than one kind of space involved. Foreground elements have the character of figure in relation to a ground; they are more distinct, more formed, and "closer" than ground, which is less distinct, less formed, and "farther away." For the terms "closer" and "farther away" one could substitute words such as "louder/softer" or "duller/brighter," but the experience is undeniably spatial, and not entirely different from the visual space of everyday life. Unless music is generated electronically and heard through headphones, any experience must also include the effects of the real space in which it is performed. The seating arrangements of the performers, size and shape of the auditorium, ray paths, reflection patterns, the directionality or lack of directionality of particular instruments, reverberation, the position of a listener vis-à-vis the musicians, all will affect the sound texture in spatial ways.

THE PERFORMING SPACE

Music as we know it is made to be performed in enclosed spaces. Instruments have a different sound in the open air; sounds blend poorly and textures seem patchy and ragged. Berlioz's blunt statement, ". . . there is no such thing as music in the open air," is right, taken in its context. The passage from his *Treatise on Instrumentation* continues:

The largest orchestra, playing in a garden open on all sides—such as the *Jardin des Tuileries*—must remain completely ineffective. Even if it were placed close to the walls of the palace, the reflection would be insufficient; the sound would be immediately lost in all directions.

An orchestra of a thousand wind instruments and a chorus of two thousand voices, placed in an open plain, would be far less effective than an ordinary orchestra of eighty players and a chorus of a hundred voices arranged in the concert hall of the Conservatoire. The brilliant effect produced by military bands in some streets of big cities confirms this statement, in spite of the seeming contradiction. Here the music is by no means in the open air: the walls of high buildings lining the street on both sides, the avenues of trees, the facades of big palaces, near-by monuments—all of these serve as reflectors. The sound is thrown back and remains for some time within the circumscribed space before finally escaping through the gaps in the enclosure. But as soon as the band reaches an open plain without buildings and trees on its march from the large street, the tones diffuse, the orchestra disappears, and there is no more music. [Berlioz, 1948:406]

The enclosed space that keeps the sound from being lost in all directions, and which is responsible for mixing, blending, and enhancing the sound, also makes it difficult for a listener to locate a particular sound source in an ensemble. Directionality is sacrificed for richness. Composers who are interested in exploiting the directionality of sound sources should keep this in mind. Gross directional effects are certainly possible, and have been used for a very long time: antiphony in cruciform churches, soloists placed in front of an ensemble, small bands of players distributed in balconies, boxes, and galleries. Such dispositions can have important effects upon textural organization. Nevertheless, directionality is often lost in the more general sensation of spaciousness, and this virtual or illusory space may be more important musically than the accurate location of sound sources.

Our ability to localize sounds is anyway rather limited unless we are able to move about, according to J. C. R. Licklider (1967). He also reminds us that localization is multisensory; that our auditory space is three dimensional and Euclidean, nearly; that sound images are not very sharply defined—auditory space is a rather diffuse thing; that auditory location persists after it has been set up; and that, if conditions are right, diverse phenomena may meld into a single percept, a single object in subjective space.

The space of stereo recording is largely an illusory space, a translation of the concert hall into a two-loudspeaker, living room situation, and some of these translations are very convincing. The engineers who place the microphones and mix the inputs are composers of this illusory and subjective space, and we are only beginning to understand how important their work is. That it is an essentially compositional effort is more apparent in the technique of multi-channel recording, where individual tracks may be laid down at different times and places. The eight- or sixteen-track master tape is capable of yielding a large number of very different final mixdowns, where the texture of the music is actually created. The space of much popular, rock, and commercial music is nondirectional, dense, and pervasive, as tactile as it is auditory. One listens with one's skin. Instead of the sound of instruments and voices in a particular place, with a translation of its spatial ambience and something of its size, the listener is submerged in a sea of sound. He swims in the ebb and swell of auditory oceanic experience, spacious and placeless.

The performing space for a rock concert is much less critical than that of a concert hall for orchestra or ensemble. Amplification and electronic reverberation can create a uniform auditory environment within almost any room. Commercial rock recordings represent the creative effort of recording engineers and producers to translate the overwhelming loudness of a live rock concert, often well above 110 db, and too often above safe sound levels for humans, into a recorded analogue which will have some power to evoke the live experience.

ARE SOUNDS SPATIAL?

Some of the remarks by Varèse about projection, sound-masses, and planes seem to be about directionality and hall characteristics, and/or about the movement of sound between loudspeakers (*Electronic Poem*); other statements imply that for him sounds were inherently spatial, and that musical space was not merely a matter of placement of sound sources, or even the physical projection of the sound through a system of loudspeakers, such as that of the Philips pavilion at the Brussels World's Fair. His remarks about his first impressions of spatial music while listening to the Beethoven *Seventh Symphony* at a concert in the Salle Pleyel are revealing:

Probably because the hall happened to be overresonant . . . I became conscious of an entirely new effect produced by this familiar music. I seemed to feel the music detaching itself and projecting itself in space. I became conscious of a third dimension in the music. I call this phenomenon "sound projection" . . . the feeling given us by certain blocks of sound. Probably I should call them beams of sound, since the feeling is akin to that aroused by beams of light sent forth by a powerful searchlight. For the ear—just as for the eye—it gives a sense of prolongation, a journey into space. [1936]

In a conversation with Gunther Schuller in 1965 he said: "I think of musical space as open rather than bounded, which is why I speak about projection in the sense that I want simply to project a sound, a musical thought, to initiate it, and then let it take its own course" (Schuller, 1965:36–37).

So projection has to do with space in the sense of standing out, and also a more general "sending forth." Varèse's musical space which is open rather than bounded is the virtual space of the music itself, independent of any particular performing space.

Musicians have paid little attention to any intrinsic spatial qualities in musical sounds and textures, but there is compelling evidence that pitch, for example, has a vertical dimension. Over forty years ago C. G. Pratt performed an experiment with clear-cut results: high tones are phenomenologically higher in space than low ones.

Observers were asked to locate on a numbered scale running from the floor to the ceiling the position of tones coming from a Western Electric No. 2-A Audiometer. The scale was $2\frac{1}{2}$ meters in height and divided into 14 equal parts. The observer sat facing the scale at a distance of three meters, while the experimenter operated the audiometer in back of a large screen to which the scale was attached. Five tones were used: 256, 512, 1024, 2048, and 4096. They were led to a telephone receiver and were presented in haphazard order at five different positions in back of the vertical scale. [1930:282]

Pratt's frequencies were chosen carefully for their octave relationships. It is well known that listeners easily confuse the octave in which a pitch is heard, and one would expect that the octave confusions would produce reversals of location. Such reversals rarely occurred. A given pitch was always placed very close to the same point. If pitches are arranged in phenomenological space according to frequency, in spite of the fact that they have been presented from different vertical positions, then pitch must be intrinsically spatial in some sense, and that was Pratt's conclusion.

A few years later S. S. Stevens, investigating tonal volume, concluded that "there is no good evidence that tones, as such, have truly spatial characteristics, and that one is justified in attributing space of extensity to auditory experience only if he is willing to accept the existence of a spatial surrogate as a sufficient basis" (1934:147). He used a congenitally blind subject, and at one point asked, "Where are the tones?" His subject replied that they were on the right arm of the chair, in a position such that she could cup her hand around them. "In order to cup her hand around the high tones, *she had to raise her hand two or three inches above the arm of the chair*" (ibid.:145; italics mine).

More recently S. K. Roffler and R. A. Butler have investigated the problem. They studied the factors that influence sound localization in the vertical plane and found that for a sound to be located accurately the stimulus must be complex (they used noise); that it must include frequencies above 7,000 Hz; and that the pinna (outer ear) must be present (1968*a*). It is well known that, while we are rather good at localizing sounds on the horizontal, we are poor at vertical sound localization, and downright bad at locating the vertical position of pure tones.

In a second paper Roffler and Butler followed the lead of Pratt in studying the vertical localization of tones (1968*b*). They used four matched loudspeakers behind a vertical panel 54 by 30 inches, divided by 4-inch strips of black cloth into thirteen numbered sections, one above the other. The centers of the loudspeakers were behind panel numbers 3, 6, 9.5, and 12. Nine frequencies that had no set of built-in relationships were used: 250, 400, 600, 900, 1400, 2,000, 3,200, 4,800, and 7,200. Listeners were positioned in different ways, and one experiment used a group of ten congenitally blind persons. The overall results supported Pratt's conclusion, that higher-pitched sounds are perceived as originating above lower-pitched ones for adults, children, and the congenitally blind. Some of their experiments showed that visual cues could influence the range of the scale within which listeners perceived the sound sources. For example, the size of the panel could influence the scale on which the tones were heard, but not the order of pitches.

The authors point out that none of their experiments is definitive with respect to whether the spatial character of tones is innate or based upon associative cues, and that such cues might be quite subtle. Nevertheless, their data supports Pratt's view that "prior to any associative addition there exists in every tone an intrinsic spatial character which leads directly to the recognition of differences in height and depth along the pitch-continuum" (1930:282).

Why, then, have composers not exploited this aspect of pitch? To my knowledge Henry Brant is the only composer who has worked with it. Of course, there was little interest in spatial aspects of music (always excepting Varèse) until the fifties, when Brant began to treat space as essential to his music. He describes an experiment where a group of performers plays a chord while arranged physically on a ladder—lowest player, lowest note, and on up, and he remarks that "the top notes seemed physically high, the bottom notes physically low" (1966:232–233).

He also describes the effect of a wall of players:

If the players can be distributed vertically from floor to ceiling, playing simultaneously in an even spread over a substantial part of the area of an entire wall, the result, especially if the instruments are arranged vertically in order of pitch (lowest notes at lowest level, etc.), will be quite as hoped for—the entire wall space will seem to be sounding at once, an extremely vivid and concentrated directional effect. [Ibid.:231]

Militating against the effect is the overwhelming strength of visual perception. We see a violinist playing and we know that the sounds, all of them, are coming from that instrument. In a darkened hall, or with our eyes closed, we might be more aware of auditory spatial effects, as Brant recommends:

The spatial elements in concert music, if exploited fully and expressively, could make their points much more strongly if the sounds could be heard in complete darkness, without the disturbing and confusing intervention of merely functional visual impressions—such as the appearance and motions of performers and audience, and the decor and lighting of the hall—that are irrelevant to the actual communication of the music in terms of its sound. [Ibid.:241–242]

Brant is likely to have his wishes come true. The advent of electronic music played through loudspeakers makes for less visual distraction—indeed, many theater pieces have come into being from a desire to give an audience some sort of visual distraction *from* the music—and opens the possibility that the intrinsic spatial properties of high and low sounds could be further developed for music.

Our ability to estimate the distance of sounds is also subject to error and confusion. Békésy set up a single loudspeaker under conditions approximating a free field, and found that

If speech sounds that had been passed through a filter were conducted to the loudspeaker, then changes in the passbands led to corresponding alterations in the perceived distance. These results show that for a correct estimation of the distance of a sound source it is necessary to have a certain amount of knowledge about its timbre. If this knowledge is absent, the listener can make a confident judgement of the distance, but this judgment will not be in agreement with the actual distance. [1960:312–313]

The other important cues to the distance of a sound source are the ratio of direct to reverberant sound and the loss of low frequencies from the sound with distance. These are, of course, sound source location cues rather than intrinsic spatial characteristics.

But the attribute of volume, the spaciousness, broadness, diffuseness of a sound, its extensity or bigness, is definitely spatial. Low tones are more voluminous, have more volume, than high tones, and most concerted music makes use of this fact. Of two tones of the same pitch, the louder will have more volume. Very high tones are packed, compact, and dense sounding, and they have correspondingly less volume. The goal of much rock music is to create a voluminous surround of sound, hence the emphasis on low bass pitches and the high level of loudness.

Békésy's (1970) paper, "Improved Musical Dynamics by Variation of Apparent Size of Sound Source," reviews older research on volume and presents new material about the apparent size of musical tones which has practical value for musicians.

Volume and the vertical orientation of pitches are inherent spatial characteristics of musical sounds, and they have important effects upon the organization of sounding textures. But positioning of instruments and sound sources such as loudspeakers can also affect texture. A string quartet with each of the players performing in a separate corner of the hall would be a very different experience than the usual compact seating arrangement. The effects of sound-source position, while very complex, are nevertheless easier to deal with than the intrinsic spatiality of sounds, so I shall review them first.

SOUND SOURCE AND TEXTURE

No composer thought more deeply about the potential of the orchestra than Berlioz. Condemned for "noisiness" and gigantism by several generations of critics dedicated to the status quo, the originality of his sound conceptions is now too often taken for granted. Two important propositions—that a composer should specify the constitution of his orchestra for each composition and that its disposition in space can be an essential part of that composition—are central to Berlioz's thinking, worked out in compositions such as the *Requiem,* and discussed in the *Treatise on Instrumentation*:

Common sense tells us that the composer—unless he is forced to employ a particular kind of orchestra—must adapt the number of performers to the character and style of his work and to the principal effects demanded by its ideas. In a *Requiem,* for instance, I have employed four small bands (trumpets, trombones, cornets, and ophicleides) placed separately at the four corners of the main orchestra, in order to render musically the monumental images of the hymn of the dead. The main orchestra consists of an imposing body of stringed instruments, of the rest of the wind instruments doubled and tripled, and of eight pairs of differently tuned kettledrums played by ten drummers. It is certain that the peculiar effects achieved by this new kind of orchestra would be impossible with any other combination.

In this connection I want to mention the importance of the different points of origin of the tonal masses. Certain groups of an orchestra are selected by the composer to question and answer each other; but this design becomes clear and effective only if the groups which are to carry on the dialogue are placed at a sufficient distance from each other. The composer must therefore indicate in his score their exact disposition. For instance, the drums, bass drums, cymbals and kettledrums may remain together if they are employed, as usual, to strike certain rhythms simultaneously. But if they execute an interlocutory rhythm, one fragment of which is given to the bass drums and cymbals, the other to kettledrums and drums, the effect would be greatly improved and intensified by placing the two groups of percussion instruments at the opposite ends of the orchestra, i.e., at a considerable distance from each other. Hence, the constant uniformity of orchestral groups is one of the greatest obstacles to the creation of monumental and truly original works. This uniformity is preserved by composers more out of habit, laziness and thoughtlessness than for reasons of economy, although the latter motive is, unfortunately, also a rather important one. [1948:407]

The plan of the "Tuba Mirum" section of the *Requiem* is such that the four bands "at the corners of the main orchestra" are not identical—those for the north and south corners are more heavily weighted toward the bass—but each contains a core of four trombones. A fifth band is comprised of the eight orchestra horns, to which four flutes, two oboes, and four clarinets are later added. The timbres of all the buzzed-lip instruments blend, the more so because of the similar character of their musical materials. There is definite and effective use of the positioning of the sound sources for their directionality. Beyond that there is the effect of spatial enlargement, which stems from the ambiguity for the listener of the actual position of any particular group—because Berlioz composes the music both within and across groups. An examination of the score will show how carefully the spatial dimension has been imagined and how wide its range. The scoring is such that it is likely to be effective in any sufficiently large performing space, because Berlioz has created a balance between groups of notoriously directional instruments which makes the most of their directionality and at the same time fully exploits their equally important power to blend.

Brant (1966) writes of a performance of the *Requiem* in Paris where the bands were at the four corners of a continuous balcony, and this disposition was used for the San Francisco performances that I heard in the early sixties. The first performance of the *Requiem* was given at the Invalides in Paris, with more than three hundred musicians grouped right and left in the transept (Barzun, 1950:277).

Brant's music for *Voyage Four*, composed in 1964, emphasizes the differences in musical materials associated with instrumental groupings. The instruments are disposed on several balconies, stage, main floor, and below. Three conductors coordinate the instrumental forces. The composer's remarks about the effects of these spatial arrangements are especially interesting.

The diagram below shows the distribution plan for my *Voyage Four*, as originally conceived, and as exactly carried out at the first performance.

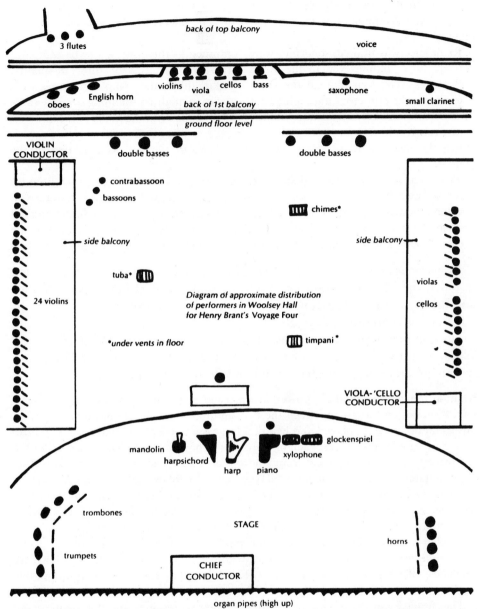

Diagram of approximate distribution of performers in Woolsey Hall for Henry Brant's Voyage Four

From "Space as an Essential Aspect of Musical Composition" by Henry Brant, in CONTEMPORARY COMPOSERS ON CONTEMPORARY MUSIC edited by Elliott Schwartz and Barney Childs. Copyright © 1967 by Elliott Schwartz and Barney Childs. Reproduced by permission of Holt, Rinehart and Winston, Inc.

At the first rehearsal the entire forces were jammed together on the stage in a conventional symphonic seating arrangement. The effects of this procedure were as follows:

> Resonance was limited, sonorities were "cramped"—there seemed to be comparatively little volume even in the large tuttis, and that seemed "thin" in quality. Balances were poor, not all the parts being audible; the complex textures were unclear—the numerous impinging contrasted tone qualities sounding over the same octave, from closely adjacent positions, produced an irritating effect of meaningless non-relationship.

When the instrumentalists later took their places in the hall as indicated in the diagram:

> There was an immediate and startling increase in volume and resonance from all the sections; heights and depths of pitch became immediately vivid; balances in volume between the superimposed but now separated textures immediately righted themselves; contrapuntal amalgams, even in the most complex places, became easily clear, and individual parts easily identifiable by direction. [1967:225–228]

In his *Gruppen für drei Orchester* (1957), Stockhausen used spatial organization in much the same way, and for much the same musical purposes, as Berlioz and Brant. The seating arrangement for the first performance, as given in the full score, is reproduced in figure 87.

Stockhausen discussed his ideas for *Gruppen* in his lecture "Music in Space." While Brant's *Voyage Four* made use of the forces usually available to a symphony orchestra, Stockhausen follows Berlioz in composing his orchestra.

> Right at the very beginning it proved necessary to present more or less long groups of sounds, noises and a cross between the two simultaneously in various tempi. So that this could be correctly played and heard, a large orchestra of 109 players was split up into three smaller orchestras: each of these was to have its own conductor and had to be placed at some distance from the other two. The three orchestras have approximately the same number of players, and the following instrumental families occur in each: woodwind, brass, plucked instruments and strings; each of these four families is again divided into a sound-group with exactly defined pitches and a noise-group with only roughly defined pitches; for the transition from sound to controlled noise within each instrumental family, a variety of metal, wood and skin percussion instruments was selected; instruments such as piano, celesta, tubular bells, cow-bells, offer a propitious connection between sound and noise when appropriately used.
>
> The similarity of the scoring of the three orchestras resulted from the requirement that sound-groups should be made to wander in space from one sounding body to another and at the same time split up similar sound-structures: each orchestra was supposed to call to the others and to give answer or echo. [1961:70]

The reawakening of interest in spatial music in our century goes back to Ives. He was wide awake in 1907 when he composed *The Unanswered Question* and other works specifying spatial separation of the instrumental forces. His

FIGURE 87

Die Sitzordnung bei der Uraufführung

Bei den folgenden Aufführungen in Donaueschingen und Wien wurden die Sitzordnungen wegen ungünstiger Saalakustik seitenvertauscht: Streicher rechts, Bläser und Schlagzeug links von den Dirigenten. Wenn möglich sollte aber die hier angegebene Aufstellung erfolgen.

Orchester II (Mitte) 36 Spieler

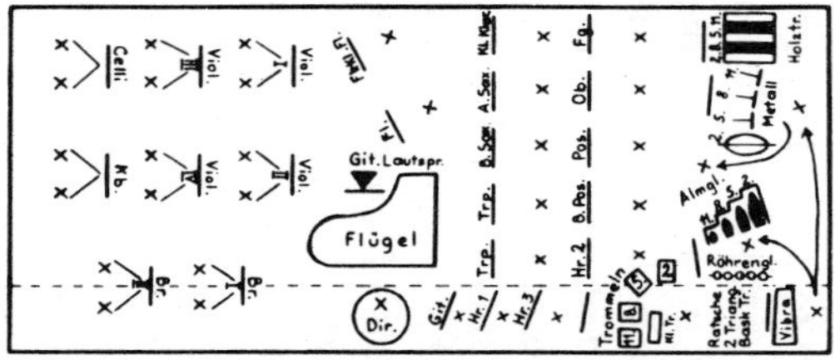

Wege für Vibraspieler zu Almglocken, Tamtam und Becken frei lassen.

Sitzordnung — im Gegensatz zur beigefügten Fotografie — etwas geändert.

Orchester I (links) 37 Spieler

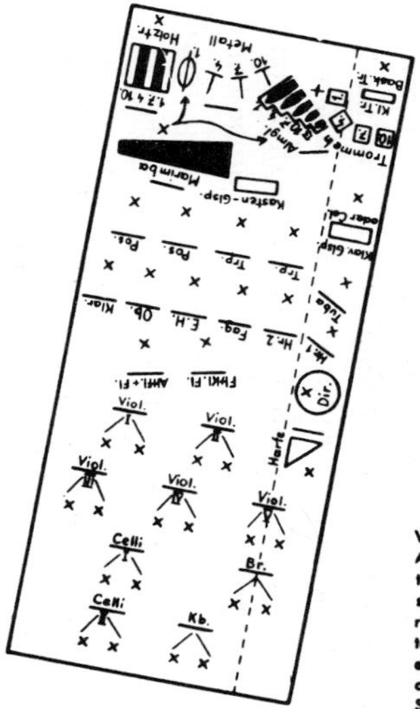

Wege für Marimbaspieler zu Almglocken, Becken und Tamtam frei lassen (Almglocken so schräg stellen, daß der Marimbaphonspieler den Dirigenten sehen kann, wenn er mit einem zweiten Schlagzeuger die Almglocken von beiden Seiten gleichzeitig schlägt). Harfe kann auch, wie in O III, am vorderen Rand des Podiums stehen.

Orchester III (rechts) 36 Spieler

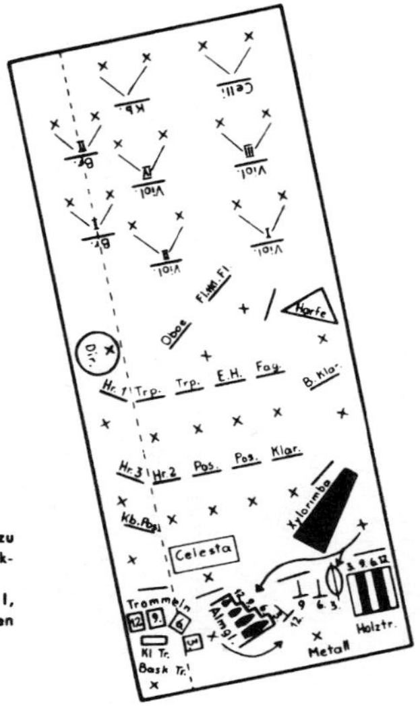

Wege für Marimbaspieler zu Almglocken, Tamtam und Becken frei lassen. Harfe kann auch, wie in O I, links neben dem Dirigenten stehen.

—— = Pult

x = Sitz oder Stand

compatriots were not. In 1929 he first published a conductor's note to the second movement of his *Fourth Symphony* which shows a remarkable sensitivity to spatial effects, both of position and texture, and real insight into problems of perspective and virtual space, which I discuss later. He views the spatial distribution of instruments as an aspect of the interpretation, and that is probably why his remarks to the conductor are general rather than specific.

To give the various parts in their intended relations is, at times, as conductors and players know, more difficult than it may seem to the casual listener. After a certain point it is a matter which seems to pass beyond the control of any conductor or player into the field of acoustics. In this connection, a distribution of instruments or group of instruments or an arrangement of them at varying distances from the audience is a matter of some interest, as is also the consideration as to the extent it may be advisable and practicable to devise plans in any combination of over two players so that the distance sounds shall travel, from the sounding body to the listener's ear, may be a favorable element in interpretation. It is difficult to reproduce the sounds and feeling that distance gives to sound wholly by reducing or increasing the number of instruments or by varying their intensities. A brass band playing pianissimo across the street is a different sounding thing than the same band playing the same piece forte, a block or so away. Experiments, even on a limited scale, as when a conductor separates a chorus from the orchestra or places a choir off the stage or in a remote part of the hall, seem to indicate that there are possibilities in this matter that may benefit the presentation of music, not only from the standpoint of clarifying the harmonic, rhythmic, thematic material, etc., but of bringing the inner content to a deeper realization (assuming, for argument sake, that there is an inner content). . . .

The writer remembers hearing, when a boy, the music of a band in which the players were arranged in two or three groups around the town square. The main group in the bandstand at the center usually played the main themes, while the others, from the neighboring roofs and verandas, played the variations, refrains, etc. The piece remembered was a kind of paraphrase of "Jerusalem the Golden", a rather elaborate tone poem for those days. The bandmaster told of a man who, living nearer the variations, insisted that they were the real music and it was more beautiful to hear the hymn come sifting through them than the other way around. Others, walking around the square, were surprised at the different and interesting effects they got as they changed position. It was said also that many thought the music lost in effect when the piece was played by the band altogether, though, I think, the town vote was about even. The writer remembers, as a deep impression, the echo part from the roofs played by a chorus of violins and voices. [1965:13]

Ives, Brant, Stockhausen, and Berlioz all make the point that by the proper positioning of instruments and instrumental groups the composer can delineate, separate, and clarify relationships between textures and their component sound elements. Brant's work was composed for a specific performing space; Berlioz and Stockhausen composed somewhat independently of any particular space; and Ives, while keenly aware of its effects, considered the management of the performing space part of the interpreter's task. Brant composed the vertical spatial dimension of pitch, Stockhausen asked for movement (wandering) of the sound between groups, and Ives drew attention to the effect of distance.

The only major idea not touched upon was that of the choric effect—probably because long acquaintance has led musicians to take it for granted. Musicians and listeners are well aware of the difference between the sound of a solo violin and a violin section playing in unison, and common sense leads one to guess that part of the little-studied choric effect is the result of small random variations in pitch, loudness, timbre, and precision of attack; but another aspect of the effect is perceptible as the size of the sound source, its dispersion, especially on the horizontal plane. Notice, for example, the placement of the strings in Brant's seating plan for *Voyage Four*; or consider the effect, in a live performance, of the passage from Berg's *Violin Concerto* where the solo violin is gradually joined by all the violins. The spatial enlargement is an important element in the music here.

Naturally, none of these spatial manifestations is reproducible in a monophonic recording. Stereophonic recordings can, on occasion, translate some spatial effects, and four-channel recording, now in home use, extends the range. The limitations of four-channel reproduction are already well known to those composers who have composed multichannel tape and electronic works for the concert hall, most importantly that the four loudspeakers remain four point sources, in the same way that the stereo pair and the monophonic speaker were point sources.

Camras made a number of experiments aimed toward recreating a sound field so that a listener could experience sound source position and ambience, to see if one could "recreate a concert hall in a telephone booth" (1968:1425). He used up to twelve microphones to define a small volume, of living room size, then played back the recorded multichannel materials in a relatively dead playback room of the same size and shape (see fig. 88). The reproduced sound gave the effect for the listener of being in a large indoor room even when reproduced in an acoustically dead space, and voices were localized quite accurately.

Unfortunately, Camras did not use musical materials, and a research group from Hitachi and Nippon Columbia (Nakayama et al., 1971) made an extensive study of the subjective effects of multichannel reproduction, following the lead of Camras, upon persons listening to light music. They used from one to eight unidirectional microphones centered on an assumed listener, all on the horizontal plane, and concluded that a four-channel system having almost the same subjective effects as one of eight channels was feasible, as long as any localization of image sources was to be in front of the listener, and that any improvement over stereo reproduction was in the ambience effect

FIGURE 88*a*. Setup for
experimental system.

FIGURE 88*b*. Location of
speakers in tests.

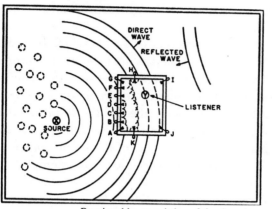

Reprinted by permission of the American
Institute of Physics and the author.

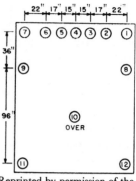

Reprinted by permission of the
American Institute of Physics
and the author.

provided by the extra speakers. They noted that other kinds of multichannel reproduction, for example, with sounds coming from many directions, presented a much more complex problem. They also found that more microphone positions improved the subjective quality somewhat, supporting Camras's results.

Much more interesting than these essentially static systems is the method developed by John Chowning (1970) for the simulation of moving sound sources. He used a powerful digital computer to generate four channels of synthesized sound. By controlling the distribution and amplitude of direct and reverberant sound among the four loudspeakers the operator is able to provide angular and distance cues. A Doppler shift enhances velocity perception. This program provides a composer of computer music with a powerful tool for accurately composing direction and position on the horizontal plane, but more important than mere position is the simulation of motion and velocity. The effects are reproducible in untreated concert halls, although the particular characteristics of the hall will naturally be superimposed upon the four channels of tape information. I have heard a demonstration of Mr. Chowning's tapes in a small concert hall with a seating capacity of about two hundred, and the effects are striking. The possibilities in composing motion are enormous, and go far beyond the "wandering" of Stockhausen's *Gruppen*. Programs such as Chowning's are certain to become an important compositional tool, especially since it appears feasible to construct analogue simulators along the same lines. For the first time it is possible for a composer to create in the studio a moving sound object and the illusory space in which it moves.

This is far more economical than the multiple-loudspeaker scheme devised by the Philips engineers (Tak, 1958–1959) for their pavilion at the Brussels World's Fair. Varèse's *Electronic Poem* was composed on three-channel tape; another synchronized fifteen-track tape carried control information for switching among sound channels through 350 spatially distributed speakers. A switching scheme was also used for the Pepsi-Cola pavilion at the 1970 Japan World's Fair.

It is the flexibility, economy, and adaptability of the Chowning program which is so attractive, compared with the high cost and fixed installation of the two pavilions. Multiple-speaker installations of some sort are nevertheless more than likely to appear with increasing frequency, and it may be possible to couple the effects of sound-source simulators with the systems of auditorium enhancement which are beginning to appear.

Peter Sutheim has reviewed several methods of actively changing hall characteristics, the most interesting of which is that of "assisted resonance," used successfully at the Royal Festival Hall in London and in the Grand Kremlin Theater in Moscow.

A very elaborate system in the enormous Grand Kremlin Theater in Moscow almost completely controls the acoustics of the hall, going far beyond mere "assistance" or "enhancement" of natural acoustics. The hall seats 6,000, so any mechanical method of controlling the acoustic properties over a broad range is impractical. Here the hall was initially made quite dead (acoustically absorbent, with a very short reverberation time). The electronic system with hundreds of speakers and microphones, augments the resonances and reverberation time in ways that are completely under the command of a skilled operator at a control console. Almost literally, the hall is "scored" or "orchestrated" along with the composition to be performed. For chamber works, a quick first reflection of sound can be fed instantly to all the speakers in the hall, creating the illusion that the room is quite small. Different parts of the hall, which would normally have different reverberant characteristics, can be brought to uniformity. Needless to say, it takes a great deal of skill to operate the system to best effect, and operators are trained especially for the job, musically as well as technically. [1969:74–75]

A room such as this, where the sense of spaciousness could be well controlled, and where the reverberation and first-reflection time could be precisely controlled on a continuous basis—in other words, composed—would be the ideal environment for the simulated movement of virtual sound sources. A new era of spatial composition is ready. The major technical problems have been solved.

BLENDING AND LAYERING

Earlier I discussed the possibility of an intrinsic spatial attribute of pitch—that high pitches seem to come from a higher spatial position. It is not so clear, in musical situations, that very low pitches seem to come from a low position—they tend to spread in all directions. Nevertheless, we do tend to perceive high and low spatially, and conventional staff notation reflects that perception.

A figure that could express the attribute of volume (extensity, bigness) would have to be somewhat different from the plane figures of staff notation. The figure below gives a sense of the voluminous lower pitches and their characteristic spread-out sound, and it expresses the compactness of the higher pitches. Any representation of volume requires some sort of three-dimensional figure.

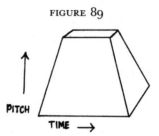

FIGURE 89

All musical textures, whether single or composite, exhibit this dimensional tendency—big on the bottom, small on top—if they have sufficient bandwidth. Some of the most interesting passages in contemporary music are those where, through timbre, dynamics, or other means, this space is restructured.

To summarize, we have, in addition to the space that may be created by the positioning of sound sources, or the movement of sound sources, an intrinsic three-dimensional pitch space. This is the space within which musical textures are constructed. Composers who work with textures directly, rather than as a by-product of other musical processes, pay careful attention to the intrinsic space of the pitch continuum.

Stratified textures express this inherent spatiality easily. We are seeing a resurgence of many techniques of layering from our own historical past, and a remarkable interest in those invented by other cultures. Much Western polyphony is stratified, from some of the earliest organa to the chorale preludes and other cantus firmus writing of Bach. Throughout a considerable portion of our history compositions were composed voice by voice, one layer after another. Mantle Hood states that "Stratification in a complete Javanese gamelan may include between thirty and forty different strata, many of these having a polyphonic relationship to one another." (1967:16); and that "Each stratum is distinguished by a particular periodicity of pitches or pulses, by its density—that is, the number of musical events occurring within an arbitrarily determined time span—by certain melodic and/or rhythmic idioms, or by a combination of these" (ibid.: 15). One needs to distinguish, however, between the techniques of construction and the sounding result. The thirty or forty layers in a gamelan composition may cohere into a single mass of sound, or a sound texture consisting of a few elements; and, in multilayered Western polyphony, one is not likely to hear all the layers unless they are strikingly contrasted in register, timbre, or rhythm.

The first master of directly composed stratified textures in this century is Debussy. Consider the opening measures of "Nuages," written in 1899 (fig. 90*a, b*).

The midrange music of the clarinets and bassoons forms a stratum. After the English horn entry the same pitches are given a different weight and quality by flutes and horns, and in measure 7 a new, higher layer of divided violins is introduced simultaneously with the low pitch of the timpani. At measure 10 an octave of violins supersedes the wider divided violins, and we hear only the thin octave on B, and distantly below, the fading sound of the timpani.

The measures from Stravinsky's *Rite of Spring*, introduction to part one, in figure 91, also present a texture, more complex, multilayered, with a few prominent melodic elements. The background is heard as a rich, rhythmized web, with E♭ clarinet and F trumpet well in the foreground.

Berg's *Altenberg Lieder*, Opus 4 (1912), often present an even denser motivic web, with little stratification or foreground/background definition of the sound elements. Particularly in the first song one hears orchestral climaxes so dense that the motivic elements are dissolved in the tide of sound. As the complexity recedes, after the second climax at measure 29, the filled-to-bursting pitch space is gradually emptied, leaving behind melodic markers that define strata. In the final measures of the first song (fig. 92), the harmonium chord, acting as a spine, is the most important element in the stratification process. The *col legno* cellos of measure 34 echo a long preoccupation with D. The motive of the three trombones is absorbed into the harmonium layer; solo violin and celeste hold an extended high E♭, while piccolo and contrabassoon touch highest and lowest boundaries. These boundaries are diffused somewhat by the "noises" of the multiple harmonic glissando of the violins and the bowed-tailpiece sound of the lower strings. The last sound heard is the harmonium chord.

Stratification is central to the musical conceptions of Charles Ives, and he uses many techniques to delineate thick, thin, chordal, contrapuntal, and rhythmic layers. His textures, even those seemingly most chaotic, are conceived as a multiplicity of strata. Moreover, he is one of a very few composers concerned with a more than conventional relationship between foreground and background elements in dense textures. His remarks to the conductor which preface the second movement of his *Fourth Symphony* are very much to the point:

As the eye, in looking at a view, may focus on the sky, clouds or distant outlines, yet sense the color and form of the foreground, and then, by bringing the eye to the

FIGURE 90*a, b*

NOCTURNES

1. N U A G E S
(Clouds)

FIGURE 91

FIGURE 92

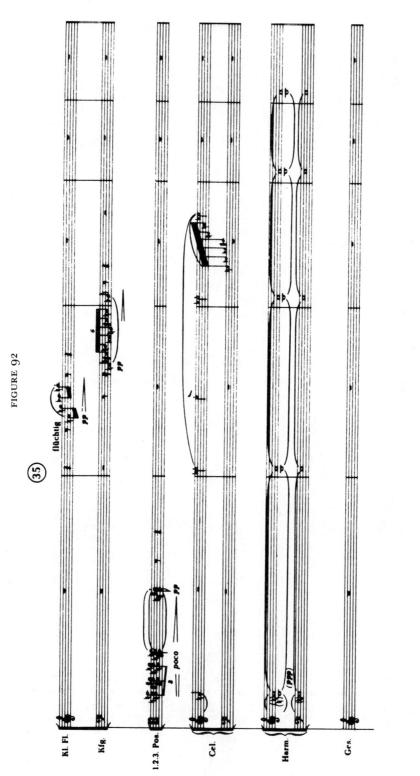

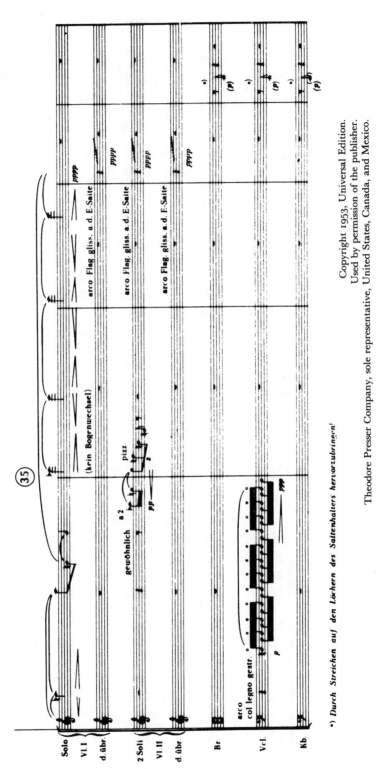

FIGURE 93

foreground, sense the distant outlines and color, so, in some similar way can the listener choose to arrange in his mind the relation of the rhythmic, harmonic and other material. In other words, in music the ear may play a rôle similar to the eye in the above instance. [1965:13]

Ives makes it clear that he is thinking of something more than a facile analogy in another paragraph:

When one tries to use an analogy between the arts as an illustration, especially of some technical matter, he is liable to get in wrong. But the general aim of the plans under discussion is to bring various parts of the music to the ear in their relation, as the perspective of a picture brings to the eye. As the distant hills, in a landscape, row upon row, grow gradually into the horizon, so there may be something corresponding to this in the presentation of music. Music seems too often all foreground even if played by a master of dynamics. [Ibid.]

Obviously, the placement of instruments may enhance the perspective aspects of the music. Ives also tried, through the actual composition of the textures, their thickness, interior dynamics, timbre, and rhythm to compose depth. He wanted to bring to the enclosed space of the concert hall the deeper and more detailed ambience of outdoor music, with its remarkable background (and sometimes middle and foreground) of "extraneous" sounds. A page from the final movement of his *Fourth Symphony* clearly shows the layering of the construction (fig. 93). The percussion group and the distant violins/harp are obvious perspective elements.

Much of the effect of the rhythmic multiplicity is not so much the separation of strata as a kind of composed haziness; the musical conception is such that each musical element contributes in some way to a sonority of depth.

FIGURE 94

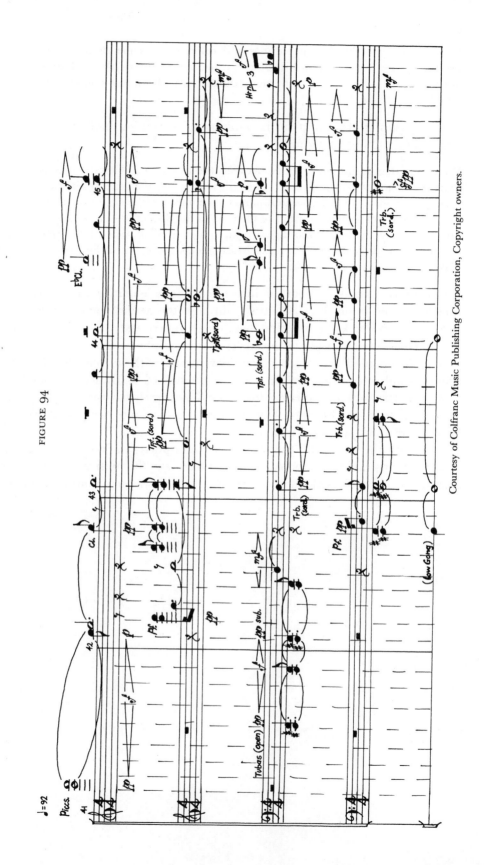

Most of the examples cited have shown fixed, stable stratification, but movement between strata (without dissolution into melodic/contrapuntal texture) is an important compositional resource. Varèse used rhythmically precise dynamics in several passages of *Déserts* to create this kind of movement (fig. 94). Within a five-octave pitch space, the layers, differentiated by pitch and instrumental quality (note the changes of instrumental color on repeated notes), are brought out individually by crescendo/diminuendo markings, precise to the eighth note. The stratification is primary, although there is strong competition from the vertical dimension, with its simple harmony of tritones and fifths. Yet the music does not sound like an arpeggiated chord; the movement is of a different order, although Varèse takes advantage of the ambiguity between the overall texture and its individual elements, including the pitches of the chord and their composite sound.

In extreme contrast with layered textures are those composite sounds which I have called fused ensemble timbres. These aim for a blend of the contributing elements in which timbral particularity is submerged in the more general sound of the whole. A fused ensemble timbre for full orchestra from the beginning of *Pli Selon Pli* by Boulez (1967) is illustrated in figure 95. At the left of the short score is the complete chord. It consists of twelve pitches, none of them doubled except soprano C and flute high C, a doubling that fleshes out the decay portion of the sound. The distribution is not unusual—large intervals at the bottom, smaller intervals at the top—except that there are some interesting clusterings and gaps. The chord is attacked *sforzando, FFFF,* by most of the instruments, and the continuation, represented by the lines, is scored for string harmonics, except for the lowest two pitches, which are played *normale.* The instrumentation of the initial attack is shown at the right. Boulez achieves here a very loud but quickly decaying sound which is

FIGURE 95

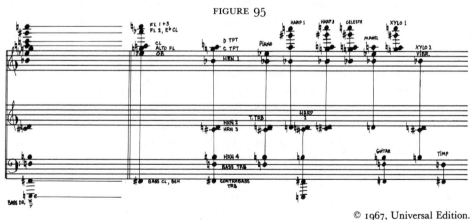

nevertheless a single perceptual and musical object, and in which a large number of discrete instrumental timbres have been dissolved. The technique is not different from the ensemble timbres of Varèse, but the number of discrete instrumental timbres employed is larger and more diversified.

The marvelous way in which Varèse used the piano to produce the attack points of ensemble timbres in *Déserts,* using only a few instruments to create extremely subtle transformations of sound, is developed by Berio in the second movement of his *Sinfonia* (1968). Piano and various brass instruments make the composite attacks, and the continuations of the sound are mixtures of voices, woodwinds, and strings. Berio has also disposed his melodic strands in such a way that one hears a banded and woven klangfarbenmelodie which is at once a melody and a texture.

The opening of the third song of Berg's *Altenberg Lieder* uses a sequence of five fused-ensemble timbres (fig. 96). There is no layering at all. Each presentation of the twelve-note chord is a single entity, different from the others only in its ensemble timbre. A note to the conductor cautions against phrasing the individual voices as melodies. Figure 97 shows the changing instrumentation of the five sounds.

There are some fascinating inner movements of the instrumental timbres: E♭ clarinet falls, trumpet and oboe rise to highest notes on the third and fourth chords, supported by rising trombone to the third chord, to create a kind of timbral climax. The B♭ clarinet rises from the second to the fifth chord, and continues a melodic figure which grows from its high B. The spacing of the chord has larger intervals on the bottom, smaller on top, without gaps, except for the tritone interval on top.

Some textures closely approach the single-object status of fused-ensemble timbres, for example, the beautiful "northern lights" chord from Ernst Krenek's *Cantata for Wartime* (1943) (fig. 98). An eleven-tone chord, in a very interesting distribution of pitches, produces a fused sound supported by a suspended cymbal roll (fig. 99).

The fused textures of the third piece of Webern's *Five Pieces for Chamber Orchestra*, Opus 10 (fig. 100), are remarkable both for the quality of the ensemble timbre created and the daring of the sound conception. The first four measures use mandolin, guitar, celeste, harp, the sounds of a single deep bell, and many cowbells to produce a combined sound that is very difficult to analyze into its constituent elements. The dynamic marking of *PPP* certainly enhances the mixing of the sounds by minimizing the effects of the attack transients—they tend to mask one another; but all those tiny transient noises then recombine into a flickering, hazy insect music, integral with the pitched elements. The sounds of the bell, less pitch specific than the other instruments, are a binding element between pitched sounds and noises. When the bass drum enters in measure 4 it adds a deep, very deep texture, a heavy distant rumble.

FIGURE 96

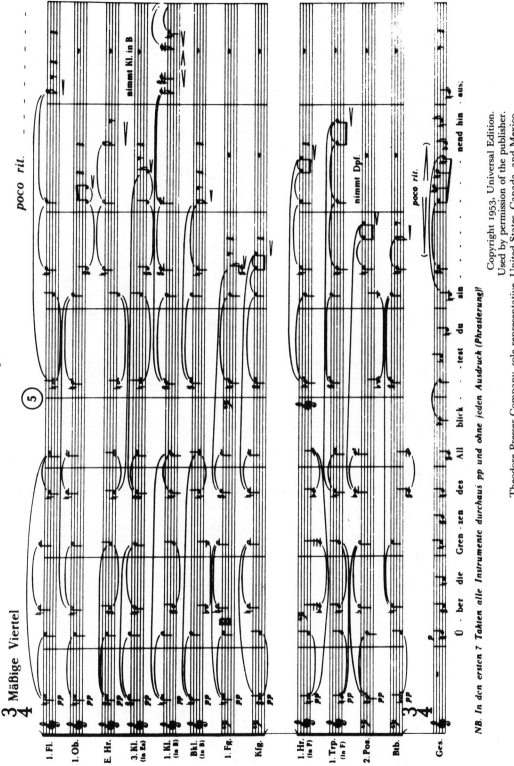

FIGURE 97

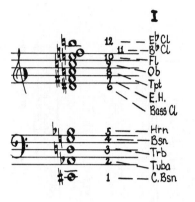

I		II	III	IV	V
12	Eb Cl	Fl	Ob	Tpt	Bb Cl
11	Bb Cl	Eb Cl	Tpt	Ob	Fl
10	Fl	Ob	Eb Cl	Bb Cl	Tpt
9	Ob	Tpt	Fl	Eb Cl	E.H.
8	Tpt	Bsn	Bb Cl	Fl	Hrn
7	E.H.	Bb Cl	Bsn	Bass Cl	Eb Cl
6	Bass Cl	E.H.	Trb	Hrn	Ob
5	Hrn	Trb	E.H.	Tuba	Bass Cl
4	Bsn	Tuba	Bass Cl	E.H.	Trb
3	Trb	C.Bsn	C.Bsn	Bsn	Tuba
2	Tuba	Bass Cl.	Hrn	C.Bsn	C.Bsn
1	C.Bsn	Hrn	Tuba	Trb	Bsn

FIGURE 98

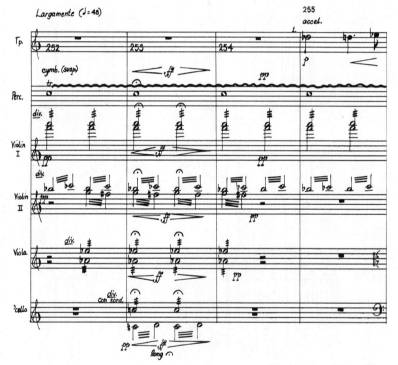

Used by permission of Ernst Krenek.

FIGURE 99

FIGURE 100

III.

FIGURE 101

At the end of the same movement (fig. 101) Webern presents another of these magical textures, this time with some changes and additions to the pitched instruments, and with a snare drum added to the percussion. The bells are used as in the opening measures, the bass drum enters the pitched texture earlier, and harmonium has displaced guitar, making for slightly less transient activity than in the opening measures. This texture is (fractionally, but this is a music of important small differences) quieter, steadier, more static than the other.

Layering and blending are rarely as distinctly characterized as in the examples I have described; much music, even music that has been composed for its textural qualities, is an amalgam of blends, layers, lines, masses, and blocks, and many compositional strategies take full advantage of the ambiguities among categories.

In *Dans le Sable*, for soprano, four altos, and small orchestra, Loren Rush has used the ambiguities of a harmonic/contrapuntal idiom to create sound that often hovers close to a music of texture. In the excerpt in figure 102 blending depends upon such things as the simple harmony, the viola drone, and the doublings between altos and instruments. The strata consist of (1) the solo violin melody, (2) the line of the altos, together with its doubling and accompaniment (but note the coloring of a single line by English horn and alto flute), (3) the texture produced by the metrically free plucked and struck instruments, and (4) the soprano solo entering at rehearsal letter Q.

Another passage (fig. 103) adds a high shimmer of divisi violins playing harmonics and harmonic glissandi, and a melodic/harmonic stratum for piano solo. Strings and woodwinds color the line of altos and flügelhorn in a manner that brings out continuities rather than the usual contrasts of klangfarbenmelodie. Beginning from K the sequence is:

1. Vln 1 and E.H.	5. Vla and E.H.
2. Vla and Alto Fl.	6. Vcl and Alto Fl.
3. Vcl and E.H.	7. Vln 1 and E.H.
4. Vln 1 and Alto Fl.	8. Vla and Alto Fl.

These timbres overlap one another, adding to the textural haze of harp, guitar, and vibraphone, and clouding the melodic/harmonic motion.

The second song of Seymour Shifrin's *Satires of Circumstance* (1964), for soprano, flute/piccolo, clarinet, violin, cello, contrabass, and piano, creates integral timbre relationships that are, however, never independent of the melodic/harmonic/contrapuntal texture (fig. 104). Timbre clearly serves these musical elements, yet it is more than a means of nuance or articulation. There is no klangfarbenmelodie in the usual sense; the timbral contrasts and continuities say, "hear the flow of the music in this way."

There is another sort of polyphony in Warren Burt's electronic composition *Moist Days in Mid-Winter* (1971). Blocks of sound coalesce and melt in slow

FIGURE 102

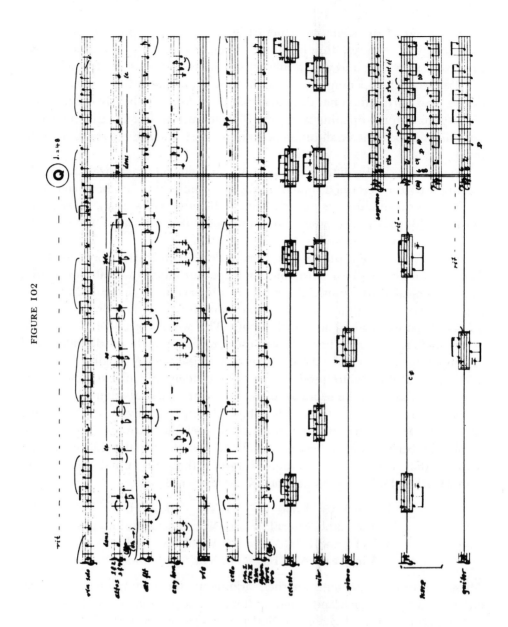

FIGURE 103

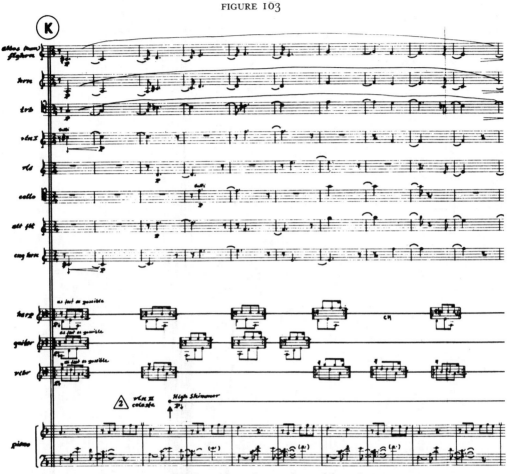

procession, sometimes opening the dense broad-band texture toward an (embedded) polyphony. There is little stratification, and the *tessitura* of the music is rather low. Movement in this composition is timbral and textural and depends upon a repertory of sounds provided by the patching. The shading of the dynamics also has much to do with the forward motion of the never quite formed "voices" of the polyphony.

Stratified textures in which there is ambiguity between the impression of layers and the sense of the texture as a whole are common enough in environmental sounds—traffic, airports, factories—but in electronic and tape music, as in music for instruments, one seldom hears layered textures of more than three perceived textural elements. Stanford Evans uses some electronically produced stratified textures in his *Each Tolling Sun*, for saxophone, dancer, and tape, which make use of microrhythm and periodicity to define and separate the layers of the pitch space. The fascination of these sounds is certainly rooted in ambiguities between the strength of the stratification and

FIGURE 104

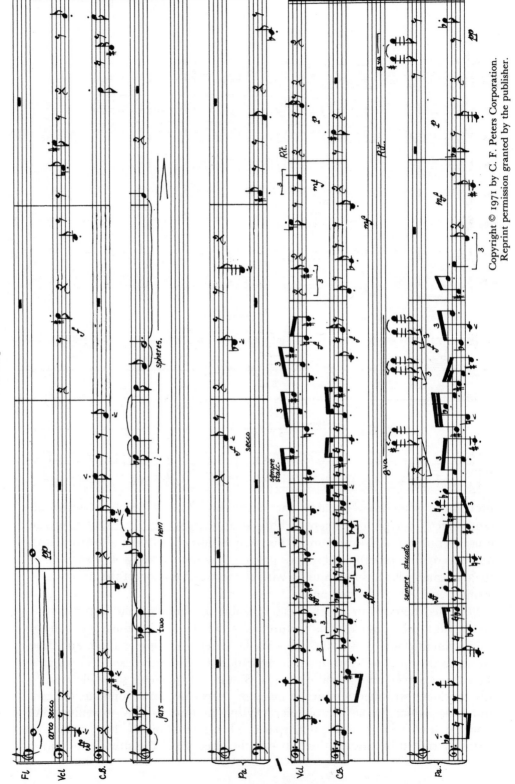

the almost equally strong "single-sound" characteristics. Only because the texture continues for an appreciable time are we able to notice and appreciate the many details of the stratification.

In my *Pacific Sirens*, for tape and instrumental ensemble, I used an ambiguous stratified texture of filtered ocean waves. A monophonic master recording of waves (from Pescadero Beach, California) was passed through band-pass filters many times to produce noise bands, which were fine tuned, remixed, and channeled. The notation in figure 105 shows the first several minutes of the pitched and reconstituted waves.

The task of the instrumentalists is to match their instrumental sounds to the pitches and quality of the performing tape and the ensemble as a whole. Melodic formations are not permitted. The players are instructed to play single pitches, corresponding to the notated tape sounds, beginning softly, swelling and diminishing, and to improvise timbre within the prescribed framework of timbre and pitch. The real notation of the music is the sound of the performing tape. In composing *Pacific Sirens* I worked at making a performing tape that not only projected the pitch strata of the band-pass filtering, but which involved internal movement—between the fixed strata, between stratified and more macroscopic textures, and between pitched and unpitched sounds. I wanted to bring out connections between those moments when the sound was an undifferentiated single thing and passages that were sharply stratified. Noise bands and noise pitches blend easily, more easily than do the pitches of most musical instruments. Players, therefore, must take care that their higher pitches—above the staff in treble clef—do not project from the textural mass. Tuning, with the tape and within the ensemble, is critical.

MASSING AND MASSES

An attractive aspect of the sound of the reconstituted waves in *Pacific Sirens* was its sense of multiplicity and its bigness. I could sometimes hear (or could imagine I could hear) the distant singing of a chorus of voices.

There is nothing new about multiplicity and the choric effect. What is new is the radical extension of the massing idea in contemporary music, and the range of its musical applications; but a great deal more needs to be known before the choric effect is fully understood or adequately synthesized. Clark et al. made some simulations, using oscillators and a recorded violin sound, and their informal tests showed "that the width of the distribution and the number of components in it were the chief factors of importance: the wider the distribution and the larger the number of partials, the more massive the chorus" (1963:53). Whether the partials were spaced randomly, or in constant or regularly increasing spacing, made little difference. Choral tones could be discriminated from solo tones at times down to 30 milliseconds. They distinguished choral from solo tones in three respects: (1) the existence of a

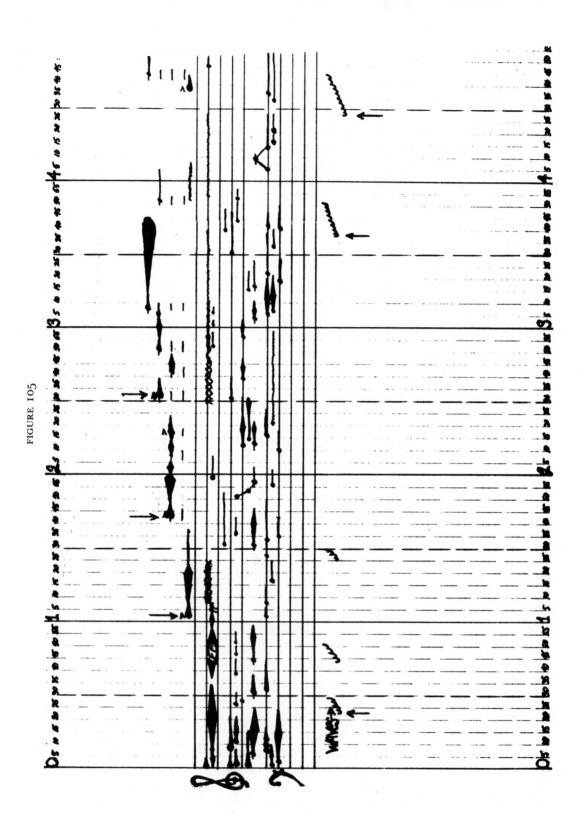

FIGURE 105

very narrow distribution of partials about each of the harmonics of a solo tone, the band being about 1 percent wide; (2) a staggered attack; (3) a spatial distribution of the instruments of the ensemble.

Fletcher and his co-workers (1963) dealt with the band of partials in a study of the quality of organ tones, and concluded that the level variation of the components of the band produced an effect of "warmth." They did not study attacks or spatial distributions. The beating among the component partials is certainly part of the choric effect, and Fletcher's "warmth" is pertinent to the richness of multiple sounds, such as a section of violins.

In the final section of his *Poesis* for Orchestra (1963), Ingvar Lidholm has extended the dimensions of the choric effect of older orchestral practice (fig. 106). A single pitch, the B♭ a seventh above middle C, undergoes a continuous change of timbre and loudness, and a most important element of the overall sound is a result of the variable vibrato of the individual instruments. There is a striking amount of fusion. Instrumental individualities are completely submerged in a slowly changing composite timbre.

Further extensions of the choric effect are certainly possible. Its use has been confined largely to voices, strings, and certain woodwinds and brasses in military and concert bands. Imagine the sound of the thirty harps of Berlioz's festival orchestra, or the effect of "combining the thirty harps with the entire mass of stringed instruments playing pizzicato into a large orchestra, thus forming a new gigantic harp with 934 strings—graceful, brilliant and voluptuous expression in all shadings" (1948:408). We have composed as though orchestras were as unchangeable as architecture. Certainly *some* of the works of the past twenty years might have profited from a disposition of instruments not so cruelly bound to older orchestral specifications. Composing should include composing the orchestra; then the orchestra can be for the composition rather than the other way round. Unless orchestral music is to go the way of the dinosaur it must have the potential for extended transformation, and no doubt this means a new form of organization. If the economics of modern life and the nature of orchestra organization is such that orchestras cannot adapt to changing musical ideas, then they will freeze into what they all too often are now—a very expensive device for accurately projecting the music of a rapidly receding past.

As an example of the effect of massing instruments other than strings consider the steel bands of the West Indies. A section of steel-band pans can produce sounds startlingly like those of an electric organ! Other equally interesting and remarkable choric transformations of our "soloist" instruments could be composed. Berlioz's festival orchestra has been criticized as the dream of a madman, but if we would take his ideas seriously—flexibility, variety, not only size—orchestral music could renew itself.

Composers of the sixties thoroughly exploited the string section of the traditional orchestra. It was the only orchestral unit large enough to be useful

FIGURE 106

UE 14420 LW

for multiple and mass effects. Penderecki's pitch-band passages can be viewed as an extension of the choric effect from a single tone to a dense band of tones, as in the passage from the end of his *Threnody to the Victims of Hiroshima* (1961) (fig. 107).

An important feature of Penderecki's pitch-band composition, and a relief from the somewhat simplistic piling up of fixed bandwidths, are the passages where the bandwidth is variable, such as at rehearsal number 10. Other important elements in this style of composition are the noise elements available from string instruments, and the important rhythmically irregular bowed tremolo and wide slow vibrato. These are essential to the detail of the sound. W. Lutoslawski discusses these microprocesses from the viewpoint of rhythm:

In reality the structure of rhythmical "collective ad libitum" is the overall sum of the rhythmical structures appearing in the particular parts. This overall sum is a phenomenon far more complex than any polyrhythmic structure appearing in traditional music. It is influenced, for example, by the possibility of the nonsimultaneous occurrence of acceleration and slowing down, in the different parts, in the performance of a section of the music. There are many similar possibilities. The result from the principle, accepted by the composer in advance, according to which each of the performers is to play, within a pre-established section of time, as if he were playing alone, not bothering about how this turns out in time in relation to the parts performed by the other members of the ensemble. In this way the rhythmical texture assumes a "flexibility" which is the characteristic feature of this music and cannot be achieved in any other way. [1968:50]

Lutoslawski's "collective ad libitum" may easily be extended to the timbral dimension. Pitch bands are hardly opaque, and all the tiny pitch differences and timbre differences produced by a large number of players make for an overall flexibility of sound, parallel to the more specifically rhythmic qualities described above. An orchestral unison is a collective ad libitum in Lutoslawski's sense, and a musical description of the choric effect is best founded upon his idea. The differences between an orchestral unison and a pitch band, even a very wide band, such as the sound at the end of Penderecki's *Threnody*, (fig. 106) are a matter of degree.

A different kind of massing is involved in cloudlike textures, but the same randomness of detail which produces the choric effect is responsible. Clouds in nature are composed of a great many droplets, and musical clouds may be composed of a great many very short sounds—pizzicato, staccato, percussive— so disposed in a time span that one hears the whole rather than any of the component elements. A cloud of this sort, complete with global characteristics, a certain pitch range and an overall shape that changes in time can be heard whenever a carbonated beverage is poured into a glass.

The transposition of the formal organization of sounds like this into musical

FIGURE 107

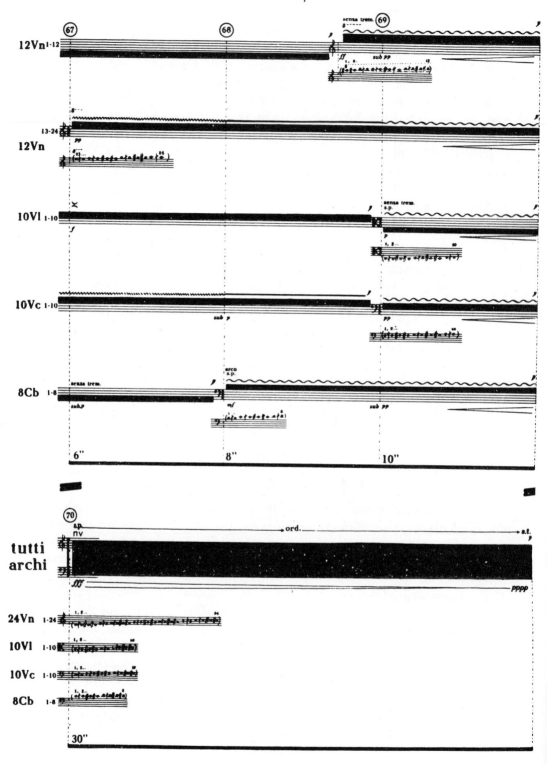

BSS 42 800
Used by permission of Belwin-Mills Publishing Corp., Melville, New York.

situations has engaged more composers during the sixties. Xenakis has written numerous works employing cloudlike textures, and his book, *Formalized Music*, has much to say about their construction:

Here then, is the technical aspect of the starting point for a utilization of the theory and calculus of probabilities in musical composition. . . . We already know that:

1. We can control continuous transformations of large sets of granular and/or continuous sounds. In fact, densities, durations, registers, speeds, etc., can all be subjected to the law of large numbers with the necessary approximations. We can therefore with the aid of means and deviations shape these sets and make them evolve in different directions. The best known is that which goes from order to disorder, or vice versa, and which introduces the concept of entropy. We can conceive of other continuous transformations: for example, a set of plucked sounds transforming continuously into a set of arco sounds, or in electromagnetic music, the passage from one sonic substance to another, assuring thus an organic connection between the two substances. To illustrate this idea, I recall the Greek sophism about baldness: "How many hairs must one remove from a hairy skull in order to make it bald?" It is a problem resolved by the theory of probability with the standard deviation, and known by the term statistical definition.

2. A transformation may be explosive when deviations from the mean suddenly become exceptional.

3. We can likewise confront highly improbable events with average events.

4. Very rarified sonic atmospheres may be fashioned and controlled with the aid of formulae such as Poisson's. Thus, even music for a solo instrument can be composed with stochastic methods.

These laws, which we have met before in a multitude of fields, are veritable diamonds of contemporary thought. They govern the laws of the advent of being and becoming. However, it must be well understood that they are not an end in themselves, but marvelous tools of construction and logical lifelines. [1971:16]

As theory this is unexceptional. Problems arise, as usual, in fitting theory to practice. Not one of Xenakis's composed clouds has the macroscopic singleness of the carbonated-drink sound mentioned above. His large numbers may not be large enough for that effect; indeed he may be composing the borderline between global sounds and their individual constituent particles. The passage from Pithoprakta (fig. 108) is a good example of a Xenakis cloud. There is a texture of *pizzicato glissando* sounds, all forty-six string instruments participating. Each instrument has a different part, and the dynamic indication is *FFF* throughout. In terms of past orchestral practice there is a very dense granulated cloud of pitches over a broad pitch range. The perceived sound is not quite so dense. Higher pitches stand out. There is a great deal of linear succession, enough so that the microlevel dominates.

If the number of individual parts could be doubled or tripled then the macrosound might begin to predominate. A dynamic marking of *PPP*, or *pizzicati* without *glissandi* would enhance the macroeffect. No criticism of Xenakis's effort is intended; presumably he made the sound he wanted to make.

FIGURE 108

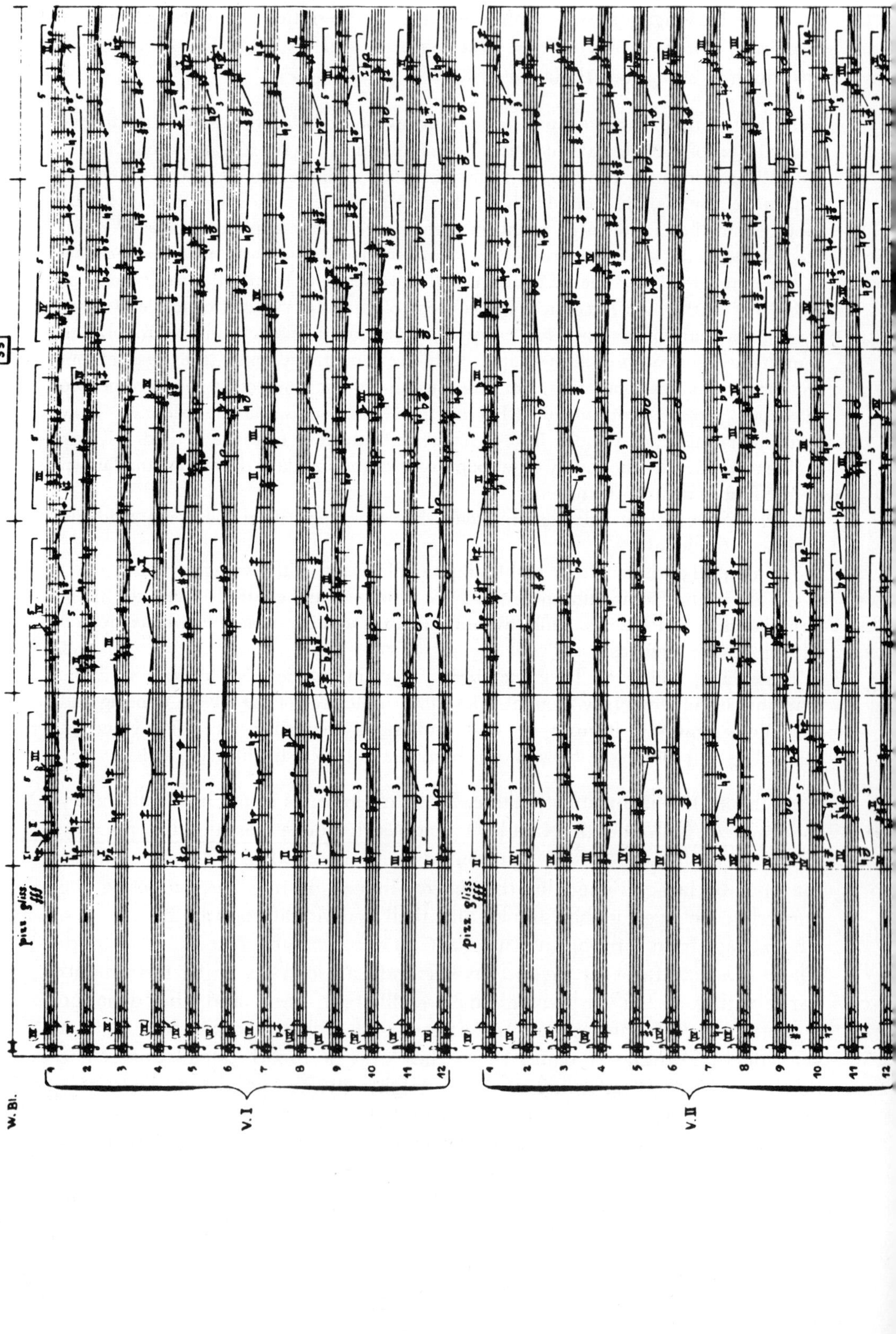

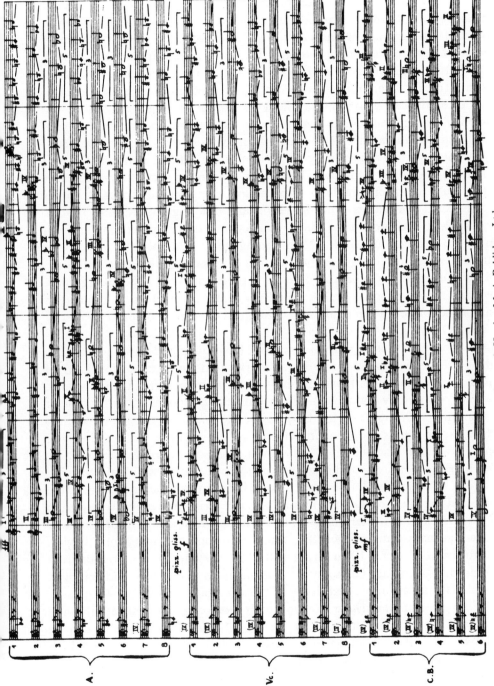

Xenakis's remarks about the control of movement within a mass, entropy as a measure of disorder, gradual or explosive deviations from a mean, improbable and average events, can certainly be applied to the composition of macroscopic events, especially if terms (this is not a small if) such as "order," "improbable," can be related precisely enough to particular sounds and musical contexts. Even a large orchestra may be too small for a wide enough range of statistical treatment, at least as far as texture and timbre are concerned. Computer sounds and textures could easily be generated, controlled and manipulated with concepts such as those that Xenakis proposes.

Movement within a mass may be accomplished without mathematics and computers, and many composers have been composing such nuanced masses by the usual handwork methods of orchestral composition. Roger Reynolds has connected two masses equal in weight and differing only in mass timbre. His *Threshold,* for orchestra, has a two-octave cluster passed intact from nonvibrato strings to winds. Maintaining the equal weight makes for continuity while the change of global timbre expresses a striking discontinuity, and the musical effect is balanced between these opposites. Schoenberg made movement between more or less equally weighted instrumental groups his basis for the subtle color modulation of the third movement of his *Five Pieces for Orchestra,* Opus 16.

Other composers, using a divisi orchestra, have composed movement within a texture. Ligeti has written several works that emphasize this kind of compositional process. All parameters—pitch, rhythm, vertical dispositions, dynamics, (especially dynamics)—contribute to the global sound. Consider the beginning of *Atmospheres* (fig. 109), where the quality of the sound changes slightly as the winds drop out imperceptibly. By measure 9 the strings are alone. Between measures 9 and 13 the lower strings move from *senza colore, non vibrato* to *poco a poco vibrato* and *poco a poco sul ponticello,* with individually varying dynamics. In measure 13 a new *tutti* texture begins, this time including all the winds, and the effect of the individually variable string dynamics (scaled from *PPP* to *FF*) is much intensified. This seamless plastic shaping of sound as texture continues until measure 23, where a melodic tremolo sound begins.

At measure 75 (fig. 110) a string texture ravels out into air sounds from buzzed-lip instruments. The sound here is meant to grow out of and complete the viola cluster, an unpitched decay of the viola pitch band.

The textures of *Atmospheres* are unusually homogeneous, and attention is therefore focused upon shadings of the global mass. A distinction can be drawn between these homogeneous masses and the kinds of texture which Varèse sometimes called sound-blocks. Sound-blocks may include many contrasting elements—short and long, loud and soft, bright and dull. The construction may be layered or not, as long as the total impression is of a single thing. Varèse's sound-blocks often remind one of different types of rocks—igneous, metamorphic, sedimentary, and so on. I have called (1965) these aggregate sound-blocks sound-icons to emphasize their single-complex-object character. Indeed, it is difficult to present two or more textures simultaneously unless they are spatially separated and/or very sharply contrasted. Ligeti has written penetratingly about these less homogeneous textures:

It is possible to distinguish various "aggregate-conditions" of the material. One can see most clearly how such conditions articulate the form in compositions where the diverse types of "weave" are accompanied by considerable differences in timbre and density, and are thus even more clearly differentiated. In Stockhausen's "Gruppen," for example, the backbone of the form is given by contrasted types—hacked, pulverized, melted, highly condensed—and their gradual transformations and mixtures one with another. In this method of composition it is vitally important to pay attention to the available degrees of permeability. The two extreme types enjoy exceptionally good mutual permeability: a dense, gelatinous, soft and sensitive material can be penetrated *ad libitum* by sharp, hacked splinters. Their mutual indifference is so great that the layers can get considerably "out" in time, and enjoy fields of inexactitude of considerable latitude. It is this peculiarity that enables the three orchestras to play together despite the fact that they are widely separated in space: the points of entry for each orchestra are generally fixed, but in the further course of a group the orchestras can diverge to a greater or lesser degree, without any damaging effect on the general result. "Soft" materials are less permeable when combined with each other, and there are places in Stockhausen's "Gruppen" of an opaque complexity beyond compare. [1965:14–15]

Certain transformations are possible within sound-blocks or aggregates: from layered to homogeneous (Varèse is the initiator); and from discrete to continuous. For the latter the steel-band sounds mentioned earlier are a simple instance of a multiplicity of discrete sounds making a perceptually continuous sound. If there were fewer pans and/or fewer note repetitions the continuity would disappear in favor of perceptually discrete details. Applying such notions to aggregate sounds should allow a composer (given a sufficiently large body of performers or electronic sound sources) to compose the middle ground between continuous and discrete.

Many composers have employed a multitude of contrapuntal voices to create a dense web of sound. The practice goes back at least to Ockeghem. Penderecki's *Threnody,* Lidholm's *Poesis,* and Ligeti's *Atmospheres* all employ multivoiced canon at some point, and each composer uses it to produce texture. Ligeti's *Atmospheres* has many passages of apparently independent voices (see fig. 111); this micropolyphony is, however, entirely in the service of the texture.

In certain passages from my *Sirens and Other Flyers* (1964) for orchestra I used up to thirty-one strings to make micropolyphonic textures. Speed (number of notes per second), pitch repetitions, pitch distributions, articulations, and dynamics are more important than melodic/contrapuntal formations, but if

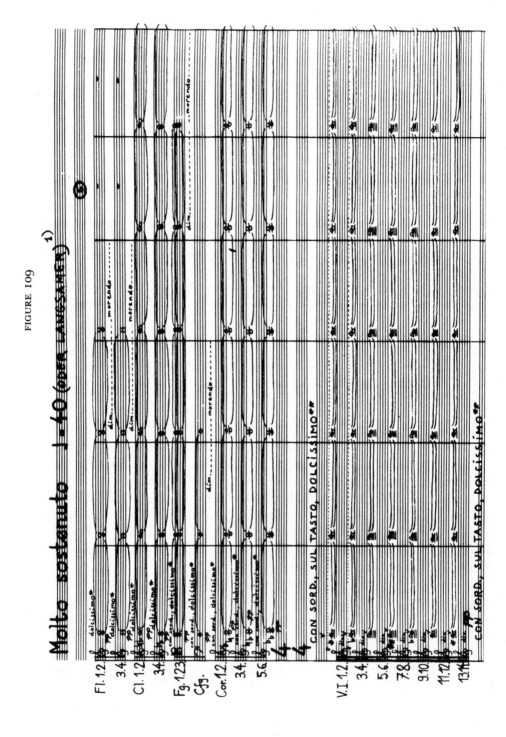

FIGURE 109

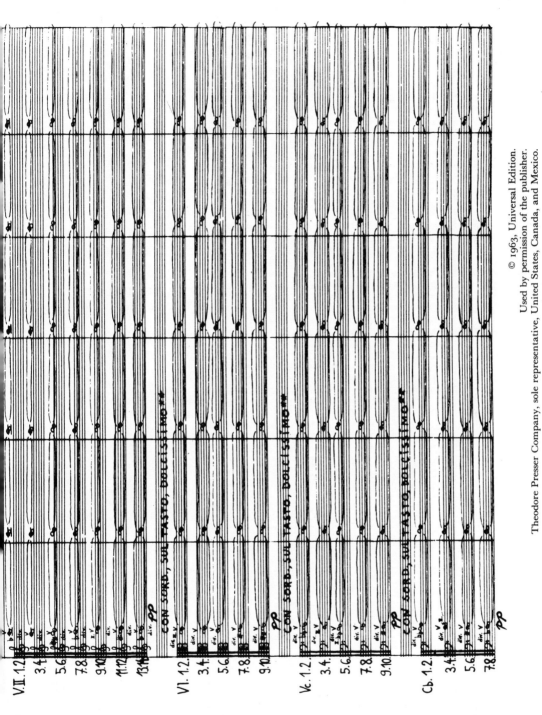

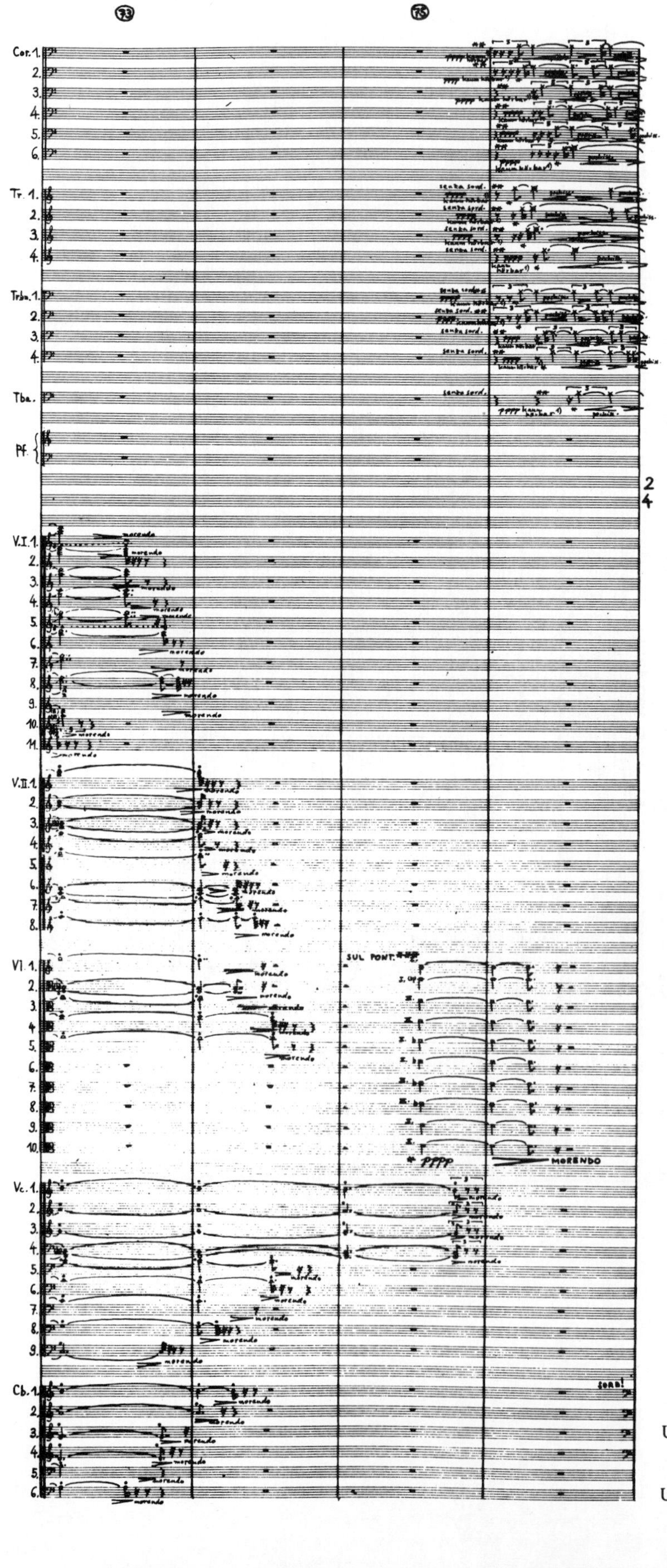

FIGURE 110

© 1963, Universal Edition.
Used by permission of the publisher.
Theodore Presser Company,
sole representative,
United States, Canada, and Mexico.

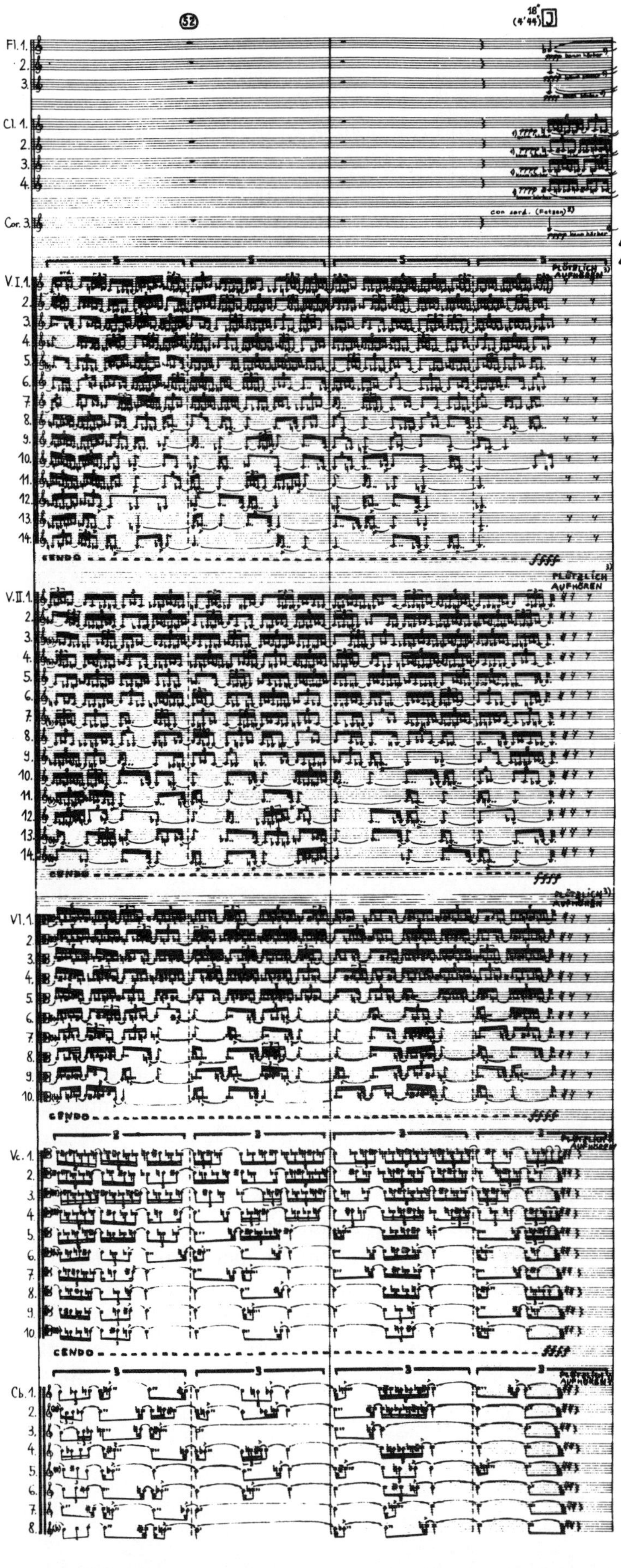

FIGURE III

the passage shown in figure 112 had been written so that individual players had parts containing only repeated notes the effect would be much different. Speed is very important here—players are often articulating at rates close to ten per second—and the different bowings contribute substantially to the textural effect.

Terry Riley's compositions acquire some of their textural qualities from speed and articulation, but his approach is melodic and contrapuntal. His *In C* (1964), for example, is built from a sequence of fifty-three short melodic motives. Each instrument goes through the sequence at a different rate, so the composition has passages where many motives are superimposed, making a very dense melodic/rhythmic web. The texture depends greatly upon the number of instruments used, and one easily available, rather dense, version (Columbia MS 7178), employs twenty-eight instruments. Melodic/contrapuntal elements are of primary significance here, but the large number of parts and the restricted pitch range of the melodic material—G below middle C to B just below soprano high C—promote the very special textural quality of this music. One can focus either on foreground elements and the detail of their rhythmic/melodic polyphony, or one can listen to the mass, taking foreground elements as bright accents in a woven pattern.

John Cage's *Atlas Eclipticalis,* discussed in the last chapter, may be heard, in some versions, as micropolyphony. The composition may be performed with any number of parts. Textural differences between versions are striking, and the sound of a full orchestra playing all eighty-six parts ought to produce a thoroughly homogenized mass of moving sound.

I have never heard a large orchestra version of *Atlas Eclipticalis,* but the recent *HPSCHD* (1967–1969) by Cage and Hiller is much concerned with the effects of massing and multiplicity. "I thought to extend this 'moving-away-from-unity' and 'moving-toward-multiplicity,' and, taking advantage of the computer facility, to multiply the details of the tones and durations of a piece of music" (1968:11).

All the sounds, including those generated by the digital computer at the University of Illinois, are related to the sounds of a harpsichord. The fifty-one channels of computer-generated sounds consist of music in equally tempered scales of from five to fifty-six pitches to the octave. The piece may be performed by from one to seven live harpsichords and one to fifty-one tapes. A recorded version (Nonesuch H-71224) uses three harpsichords, one electric and one with a 17 percent time compression, and a mixdown of the fifty-one channels of sound. The harpsichordists play musical fragments from works by Mozart and later composers, sometimes with left- and right-hand parts disassociated. Listening to the recorded version is an amazing experience, full of surprises and perceptual kinks. Tunes and other coherent borrowed musical materials slip in and out of perception, their beginning and ending points

FIGURE 112

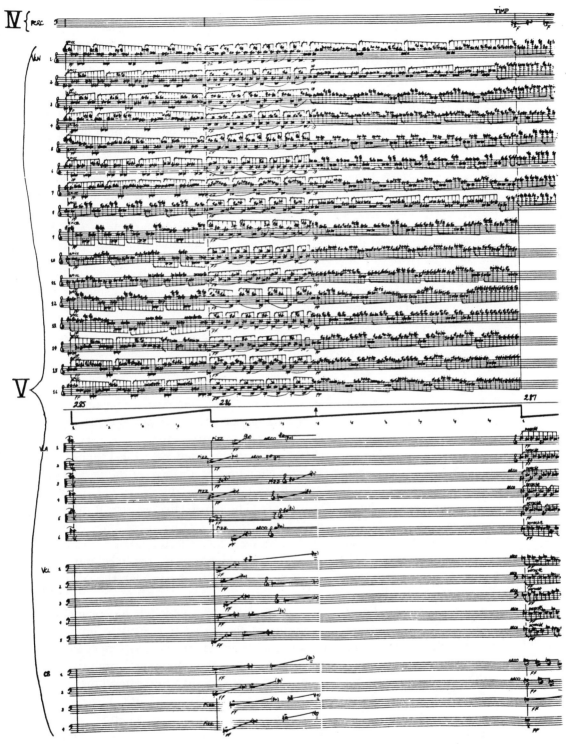

masked by the multitude of less-patterned sounds. It is like a walk through a forest. One sees the pine among the maples or the maple among the pines, and sometimes, if there are many varieties, one sees only the forest.

HPSCHD is full of this kind of perspective. It requires a level of multiplicity—a sprawling jungle of sound—which would be hard to make without the help of a computer. What is startling and instructive is the depth of perspective in the mass of sound. The wide range of micropitch organizations and the amount of timbral variation have much to do with this:

We have tried—Jerry Hiller and I—to give a quality of fine division not only to the pitches but to the durations and also to the timbre, which will be, in general, imitative of harpsichord sound: an attack followed by a decay that has an inflection point. The decay is not a straight line but a line with a bend in it. It starts down and then continues at a different descending angle. Now, that inflection point can be moved and the angle changed and so give "microtimbral" variations, and *that* we've related to the chart of the I Ching. The first "subroutine" we made for the computer was to substitute for manual tossing of coins—to obtain the numbers one to sixty-four. This subroutine was used in order to find at which point this inflection—this change of decay, this place in the sound—changes from note to note. It ought to be, in the end—and as I told you today, we haven't heard a single sound—not only micro-tonal and micro-durational, but micro-timbral. [Ibid.]

HPSCHD could not work without the history of music and our memories of that music. The fragments that we recognize as coherent patterns and which we try to follow are largely responsible for the striking depth and perspective in the textural mass. These recognizable bits are constantly emerging and submerging in a rather mysterious way, and the listener may feel that *he* is in motion, because these experiences of changing perspective of the details in a mass are so much a part of modern life, where we see things (and hear things) from moving automobiles and airplanes. The perspective aspects of the piece can be grasped easily if one plays *HPSCHD* one channel at a time. Channel two has a great many easily recognizable elements of historical music; channel one employs a finer shredder, but occasional fast scale passages and other coherent bits emerge. Both channels at once are needed for the complete walk-through-the-forest.

All dense masses have some sort of perspective which is independent of sound source position, but positioning of the sound sources is so important (and the thicker the mass the more important it is) that new, more flexible performance environments must be devised. The architectural problems can only be solved when architects and acoustical engineers understand the *musical* import of sound-source position and sound perspective. Until new physical enclosures are devised we must be satisfied with multichannel speaker installations, the assisted resonance described earlier in this chapter, and the Chowning method of sound movement.

These ideas about perspective, masses, and massing are not particularly recent. Forests of sound were dreamed by Mozart, Beethoven, Berlioz, Wagner, and many others—think of opera and oratorio right from their beginnings—long before the composition of *HPSCHD*. And rather early in our century Ives outlined these conceptions in their modern form:

When we were in the Keene Valley, on the Plateau, in 1915, with Edie, I started something that I had had in mind for some time: trying out a parallel way of listening to music suggested by looking at a view.

First, with the eyes toward the sky or tops of the trees, taking the earth or foreground subjectively (that is, not focussing the eye on it), and then

Second, looking at the earth and land and seeing the sky and the top of the foreground subjectively. In other words, giving a musical piece in two parts, but both played at the same time . . . the whole played through twice, first when the listener focusses his ears on the lower or Earth music, and the next time on the upper, Heavens music. . . .

The earth part is represented by lines, starting at different points and at different intervals. A kind of uneven and interlapping counterpoint, sometimes reaching nine or ten different lines representing the ledges, rocks, weeds and land formations, lines of trees and forest, meadows, roads, rivers, and undulating lines of mountains in the distance, that you catch in a wide landscape. And with this counterpoint, a few of the instruments playing the melodic lines are put into a group, playing masses of chords built around intervals, in each line. This is to represent the body of the earth, where the rocks, trees and mountains arise. Between the lower group and the upper, there is a vacant space of four tones between B-natural and E-natural.

The part of the orchestra representing the Heavens has its own chord system, but its counterpoint is chordal. . . . There are three groups in some places divided into four or five. On the lower corner of the second page of the sketch, this chordal counterpoint is broken by long chords, but stays this way for only a short time. These two main groups come into relation harmonically only in cycles, that is, they go around their own orbit and come to meet each other only where their circles eclipse. [Cowell, 1955:201–202]

For performance of this *Universe Symphony* Cowell (1955) writes that Ives imagined several different orchestras, with huge conclaves of singing men and women, placed in valleys, on hillsides, and on mountains. He never meant to finish it, and thought of this work as a possible collaborative effort by several composers.

It is too late for Ives to hear the sounds of the music described above, but the ideas underlying that vision, multiplicity, sound-source groupings, sound perspective, sound as embedded in the natural order, sound in its physicality and its endless variety, undergird much of the significant music we are hearing today.

Bibliography

American Standards Association

    1960. *American Standard Acoustical Terminology*. New York. Definition 12.9, Timbre, p. 45.

Babbitt, Milton

    1965. "The Electronic Synthesis and Perception of Music." Lecture, March 13, University of Illinois, Urbana. Tape recording.

Backhaus, S.

    1932. "Über die Bedeutung der Ausgleichsvorgänge in der Akustik." *Zeitschrift für technische Physik*, 13, 1:31–46

Barzun, J.

    1950. *Berlioz and the Romantic Century*. Boston, Little Brown and Co.

Beauchamp, J. W.

    1966. "Additive Synthesis of Harmonic Musical Tones." *Journal of the Audio Engineering Society*, 14, 4:332–342.

    1972. Letter, *DB Magazine*, 6, 7:4.

Békésy, G.

    1960. *Experiments in Hearing*, trans. and ed. E. G. Wever. New York, McGraw-Hill.

    1970. "Improved Musical Dynamics by Variation of Apparent Sound Source." *Journal of Music Theory*, 14, 3:142–164.

    1971. "Auditory Backward Inhibition in Concert Halls." *Science*, 171, 3971:529–536.

Berlioz, Hector

    1948. *Treatise on Instrumentation*, ed. Richard Strauss, trans. Theodore Front. New York, Edwin F. Kalmus.

Bowra, C. M.

    1962. *Primitive Song*. Cleveland and New York, World Publishing Co.

Brant, Henry

    1966. "Space as an Essential Aspect of Musical Composition." In *Contemporary Composers on Contemporary Music*, ed. Elliott Schwarz and Barney Childs. New York, Holt, Rinehart and Winston. Pp. 223–242.

Bregman, Albert S.
 1971. "On the Perception of Sequences of Tones." Paper presented at American Association for the Advancement of Science, New York. Mimeo.
Bregman, Albert S., and Jeffrey Campbell
 1970. "Auditory Channelization and Perception of Order in Rapid Sequences of Tones." McGill University, Toronto. Mimeo.
 1971. "Primary Auditory Stream Segregation and Perception of Order in Rapid Sequences of Tones." *Journal of Experimental Psychology*, 89, 2:244–249.
Bregman, Albert S., and G. L. Dannenbring
 1973. "The Effect of Continuity on Auditory Stream Segregation." *Perception and Psychophysics*, 13, 2:308–312.
Brown, Robert E.
 1965. "The Mṛdaṅga: A Study of Drumming in South India." 2 vols. Ph.D. dissertation, University of California, Los Angeles. Ann Arbor, University Microfilms.
Busoni, F.
 1911. "Sketch for a New Esthetic of Music." In *Three Classics in the Aesthetic of Music*, trans. Theodore Baker. New York: Dover Publications, 1962.
Cage, John
 1962. *Catalogue of Works.* New York, Henmar Press.
Cage, John, and Lejaren Hiller
 1968. "HPSCHD." *Source Magazine*, 2, 2:10–19
Camras, Marvin
 1968. "Approach to Recreating a Sound Field." *Journal of the Acoustical Society of America*, 43, 6:1425–1431.
Carterette, Edward C., Morton P. Friedman, and John D. Lovell
 1968. "Mach Bands in Hearing." *Journal of the Acoustical Society of America*, 45, 4:986–998.
Chou Wen-chung
 1966. "The Liberation of Sound: Edgar Varèse." *Perspectives of New Music*, 5, 1:1–19.
Chowning, John M.
 1970. "The Simulation of Moving Sound Sources." Audio Engineering Society Convention Preprint 726 (M–3). Mimeo.
Clark, Melville, Jr., et al.
 1963. "Preliminary Experiments on the Aural Significance of Parts of Tones of Orchestral Instruments and on Choral Tones." *Journal of the Audio Engineering Society*, 11, 1:45–54.
Cowell, Henry, and Sidney Cowell
 1955 *Charles Ives and His Music.* New York, Oxford University Press.
Crossley-Holland, Peter
 1954. "Tibetan Music." In *Grove's Dictionary of Music*, 5th ed. New York, St. Martin's Press. P. 459.
 1970. "The Music of the Tantric Rituals of Gyume and Gyutö." Record booklet for AST-4005. New York, Anthology Record and Tape Corporation. Pp. 3–7.

Denes, P. B.
 1963. "On the Statistics of Spoken English." *Journal of the Acoustical Society of America*, 35, 6:892–904

Ellingson, Terry
 1970. "The Technique of Chordal Singing in the Tibetan Style." *American Anthropologist*, 72, 4:826–831.

Ellis, C. J.
 1964. *Aboriginal Music Making.* Adelaide, Libraries Board of Southern Australia.

Erickson, Robert
 1965. "Varèse: 1924–1937: Sound Ikon." *Nutida Musica*, 4, 6:18–22. In Swedish. Reprinted in *The Composer*, 1, 3 (1969), 144–149.

Fletcher, Harvey, E. Donnel Blackham, and Richard Stratton
 1962. "Quality of Piano Tones." *Journal of the Acoustical Society of America*, 34, 6:749–761.

Fletcher, Harvey, Donnell E. Blackham, and Douglas A. Christensen
 1963. "Quality of Organ Tones." *Journal of the Acoustical Society of America*, 35, 3:314–325.

Fletcher, Harvey, and L. Sanders
 1967. "Quality of Violin Vibrato Tones." *Journal of the Acoustical Society of America*, 41, 6:1534–1544.

Greenwalt, C.
 1968. *Bird Song: Acoustics and Physiology.* Washington, Smithsonian Institution Press.

Griffin, D.
 1968. "Echolocation and Its Relevance to Communications Behavior." In *Animal Communication*, ed. Thomas A. Sebeok. Bloomington, Indiana University Press. Pp. 155–164.

Helmholtz, Hermann
 1857. "On the Physiological Causes of Harmony in Music," trans. A. J. Ellis, in *Popular Scientific Lectures*, selected and introduced by Morris Kline. New York: Dover Publications, 1962. Pp. 22–58. A lecture delivered in Bonn in winter of 1857.
 1877. *On the Sensations of Tone As a Physiological Basis for the Theory of Music.* 2d English ed., to conform to 4th German ed. of 1877. Trans. Alexander Ellis, intro. by Henry Margenau. New York: Dover Publications, 1954.

Hickman, H., and Charles Gregoire, Duc de Mecklembourg
 1958. *Catalogue d'Enregistrements de Musique Folklorique Egyptienne.* Strasbourg and Baden-Baden, P. H. Heitz.

Hood, Mantle
 1966. "Slendro and Pelog Redefined." In *Selected Reports*, 1,1. Institute of Ethnomusicology, University of California, Los Angeles. Pp. 28–48.

Hood, Mantle, and Hardja Susilo
 1967. *Music of the Venerable Dark Cloud*, Los Angeles, Institute of Ethnomusicology, University of California, Los Angeles (with accompanying record).

Ives, Charles
 1965. Conductor's note, pp. 12–14, of performance score, Symphony No. 4,

Associated Music Publishers, New York, 1965. Reprinted from *New Music Edition*, January 1929.

Jones, Trevor A.
  1956. "Arnhem Land Music, Pt. II, A Musical Survey." *Oceania*, 26, 4:252–339.
  1957. "Arnhem Land Music, Pt. II, A Musical Survey (continued)." *Oceania*, 28, 1:1–30.
  1965. "Australian Aboriginal Music: The Elkin Collection's Contribution toward an Overall Picture." In *Aboriginal Man in Australia: Essays in Honour of Emeritus Professor A. P. Elkin*, ed. R. M. and C. H. Berndt. Sydney, Angus and Robertson.
  1967. "The Didjeridu: Some Comparisons of Its Typology and Musical Functions with Similar Instruments throughout the World." *Studies in Music* (University of Western Australia Press), 1:23–55.

Karlin, J. E.
  1945. "Auditory Tests for the Ability To Discriminate the Pitch and Loudness of Noise." ORSD Report 5294.

Kaufmann, Walter
  1967. *Musical Notations of the Orient*. Bloomington and London, Indiana University Press.

Kircher, Athanasius
  1650. *Musurgia Universalis*, ed. Ulf Scharlau. Hildesheim and New York: G. Olms, 1970. Reprint of the 1650 edition, Rome.

Kirk, R.
  1959. "Tuning Preferences for Piano Unison Groups." *Journal of the Acoustical Society of America*, 31, 12:1644–1648.

Kolers, Paul A.
  1968. "Some Psychological Aspects of Pattern Recognition." In *Recognizing Patterns: Studies in Living and Automatic Systems*, ed. Paul A. Kolers and Murray Eden. Cambridge, Massachusetts Institute of Technology Press.

Kolneder, W.
  1961. *Anton Webern: Einführung in Werk und Stil*. Rodenkirchen and Rhein, P. J. Tonger.

Lenneberg, Eric H.
  1960. "Language, Evolution and Purposive Behavior." In *Culture in History: Essays in Honor of Paul Radin*, ed. S. Diamond. New York, Columbia University Press. Pp. 869–893.

Liberman, A. M.
  1970. "The Grammars of Speech and Language." *Cognitive Psychology*, 13:301–323.

Licklider, J. C. R.
  1967. "Phenomena of Localization." In *Sensorineural Hearing Processes and Disorders*, ed. A. Bruce Graham. Boston, Little Brown and Co. Pp. 123–127.

Ligeti, György
  1965. "Metamorphoses of Musical Form." *Die Reihe*, 7:5–19

Lorenz, K.
  1961. "The Role of Gestalt Perception in Animal and Human Behaviour." In

Aspects of Form, ed. Lancelot Law Whyte. Bloomington, Indiana University Press. Pp. 157–178.

Luce, David A.
　1963. "Physical Correlates of Nonpercussive Musical Instrument Tones." Ph.D. dissertation. Massachusetts Institute of Technology, Cambridge, Mass.

Luce, D., and Melville Clark, Jr.
　1965. "Durations of Attack Transients of Nonpercussive Orchestral Instruments." *Journal of the Audio Engineering Society*, 13, 3:194–199.

Lutoslawski, Witold
　1968. "About the Element of Chance in Music." In *Three Aspects of New Music*, György Ligeti, Witold Lutoslawski, and Ingvar Lidholm. Stockholm, Nordiska Musikförlaget. Pp. 47–53.

Martino, D.
　1966. "Notation in General—Articulation in Particular." *Perspectives of New Music*, 4, 2:47–58

Melka, A.
　1970. "Messungen der Klangeinsatzdauer bei Musik-instrumenten." *Akustika*, 23,2: 108–117.

Miller, G. A.
　1962. "Recognizing and Identifying." In *Psychology: The Science of Mental Life*. New York, Harper and Row. Pp. 147–159.
　1965. "Some Preliminaries to Psycholinguistics." American Psychologist, 20:15–20. Repr. in *Language*, ed. R. C. Oldfield and J. C. Marshall, Penguin Books, Ltd. 1968, pp. 202–212.

Miller, G. A., and G. A. Heise
　1950. "The Trill Threshold." *Journal of the Acoustical Society of America*, 22, 5:637–638.

Nakayama, T., et al.
　1971. "Subjective Assessment of Multichannel Reproduction." *Journal of the Audio Engineering Society*, 19, 9:744–751.

Nordmark, Jan O.
　1970. "Time and Frequency Analysis." In *Foundations of Modern Auditory Theory*, Vol. I, ed. Jerry V. Tobias. New York and London, Academic Press. Pp. 57–83.

Patterson, J. H., and D. M. Green
　1970. "Discrimination of Transient Signals Having Identical Energy Spectra." *Journal of the Acoustical Society of America*, 48, 4 (part 2):894–905.

Payne, R.
　1970. *The Whale*. Solana Beach, Calif., CRM Books.

Picken, L.
　1957. "The Music of Far Eastern Asia. 1. China." In *Ancient and Oriental Music, The New Oxford History of Music*, ed. Egon Wellesz. London, Oxford University Press. Pp. 83–134

Piston, W.
　1947. "Strawinsky's Rediscoveries." *Dance Index*, 6, 10/11/12:256–257.

Plomp, Reinier
　1964. "The Ear As a Frequency Analyzer." *Journal of the Acoustical Society of America*, 36, 9:1628–1636

1966. *Experiments on Tone Perception*, Soesterberg, Institute voor Zintuigfysiologie RVO–TNO.

Pousseur, Henri

1968. "Calculation and Imagination in Electronic Music." *Electronic Music Review* 5, Jan.:21–29

Pratt, C. C.

1930. "The Spatial Character of High and Low Tones." *Journal of Experimental Psychology*, 13, 4:278–285.

Rainbolt, Harry R., and Earl D. Schubert

1968. "Use of Noise Bands To Establish Noise Pitch." *Journal of the Acoustical Society of America*, 43, 2:316–323.

Risset, J. C.

1970. "An Introductory Catalog of Computer Synthesized Sounds." Murray Hill, N.J., Bell Telephone Laboratories. Mimeo with recorded examples. No. 550, p. A–103.

Rock, J. F.

1940. "The Romance of ² K'a-² Mä-²Gyu-³ Mi-²Gkyi." *Bulletin de L'Ecole Francaise d'Extreme-Orient*. Hanoi. 39:1–152.

Roffler, S. K., and R. A. Butler

1968a "Factors That Influence the Localization of Sound in the Vertical Plance." *Journal of the Acoustical Society of America*, 43, 6:1255–1259.

1968b "Localization of Tonal Stimuli in the Vertical Plane." *Journal of the Acoustical Society of America*, 43, 6:1260–1266.

Rufer, J.

1969. "Noch einmal Schönbergs Op. 16." *Melos*, 9:366–368.

Saldanha, E. L., and J. F. Corso

1964. "Timbre Cues and the Identification of Musical Instruments." *Journal of the Acoustical Society of America*, 36, 11:2021–2026

Sayre, K. M.

1968. "Toward a Quantitative Model of Pattern Formation." In *Philosophy and Cybernetics*, ed. Frederick J. Crosson and Kenneth M. Sayre. New York, Simon and Schuster. Pp. 149–152.

Schoenberg, A.

1911. *Harmonielehre*. Leipzig and Vienna, Universal Edition no. 3370.

Schouten, J. F.

1940. "The Residue: A New Component in Subjective Sound Analysis." *Proceedings, Koninklijke Nederlandsche Akademie van Wetenschappen*, 43, 3:356–365.

1962. *The Perception of Sound and Speech*, 4th International Congress on Acoustics Reports II, Copenhagen. Pp. 195–205.

1968. *The Perception of Timbre*. Reports of the 6th International Congress on Acoustics, Tokyo, GP-6-2. Pp. 35–44, 90.

Schouten, J. F., R. J. Ritsma, and B. Lopes Cardozo

1962. "Pitch of the Residue." *Journal of the Acoustical Society of America*, 34, 8 (part 2):1418–1424.

Schuck, O. H., and R. W. Young

1943. "Observations on the Vibrations of Piano Strings." *Journal of the Acoustical Society of America*, 15, 1:1–11.

Schuller, G.

1965. "Conversation with Varèse." *Perspectives of New Music*, 3, 2:32–37.

Simon, H. A.

1969. "The Architecture of Complexity." In *The Sciences of the Artificial*, Cambridge, Mass., Massachusetts Institute of Technology Press. Pp. 84–118. Reprinted from *Proceedings of the American Philosophical Society*, 106 (Dec. 1962), 467–482.

Slawson, A. Wayne

1965. *A Psychoacoustic Comparison between Vowel Quality and Musical Timbre*, Fifth International Congress of Acoustics, Liege. M66.

1968. "Vowel Quality and Musical Timbre As Functions of Spectrum Envelope and Fundamental Frequency." *Journal of the Acoustical Society of America*, 43, 1:87–101.

Slaymaker, F. H.

1970. "Chords from Tones Having Stretched Partials." *Journal of the Acoustical Society of America*, 47, 6:1569–1571.

Smith, Huston, Kenneth N. Stevens, and Raymond S. Tomlinson

1965. "On an Unusual Mode of Chanting by Certain Tibetan Lamas." *Journal of the Acoustical Society of America*, 41, 5:1262–1264.

Stanford, W. B.

1967. *The Sound of Greek*. Berkeley and Los Angeles, University of California Press.

Stevens, S. S.

1934. "Are Tones Spatial?" *American Journal of Psychology*, 46:145–147.

Stockhausen, K.

1964. Interview with Eric Salzman, WBAI, New York.

1961. "Music in Space." *Die Reihe*, 5:67–82.

Stravinsky, I., and R. Craft.

1963. *Dialogues and a Diary*. New York, Doubleday and Co.

Studdert-Kennedy, Michael, and Donald Shankweiler

1970. "Hemispheric Specialization for Speech Perception." *Journal of the Acoustical Society of America*, 48, 2 (part 2):579–594.

Sutheim, Peter

1969. "Electro-Acoustics in the Concert Hall." *Stereo Review*, April:72–76

Szöke, P., et al

1969. "The Musical Microcosm of the Hermit Thrush." *Studia Musicologica Academiae Scientiarum Hungaricae*, 11:423–438.

Tak, W.

1959. "Electronic Poem." Philips Review, 2/3:43–49.

Tenney, James C.

1965. "The Physical Correlates of Timbre." *Gravesaner Blätter*, 7, 26:106–109.

1969. "Computer Music Experiments, 1961–1964." *Electronic Music Reports* (Institute of Sonology, Utrecht), 1:23–60.

Thomas, I. B.

1969. "Perceived Pitch of Whispered Vowels." *Journal of the Acoustical Society of America*, 46, 2 (part 2): 468–470.

Van Gulick, R. H.

    1969. *The Lore of the Chinese Lute.* Tokyo and Rutland, Vt., C. E. Tuttle.

    1969. *Hsi K'ang and His Poetical Essay on the Lute.* Tokyo and Rutland, Vt., C. E. Tuttle.

Varèse, Edgard

    1936. *New York Times*, Dec. 6, 1936, Section L, Part N, p. 7.

Ward, W. Dixon

    1970. "Musical Perception." In *Foundations of Modern Auditory Theory*, Vol I., ed. Jerry V. Tobias. New York and London, Academic Press. Pp. 407–447.

Warren, Richard M., and Roslyn P. Warren

    1968. *Helmholtz on Perception: Its Physiology and Development.* New York, Wiley and Sons.

Weidlé, W.

    1957. "Biology of Art: Initial Formulation and Primary Orientation." *Diogenes*, 17, Spring:1–15.

Weiss, P.

    1969. "The Living System: Determinism Stratified." In *Beyond Reductionism*, ed. A. Koestler and J. R. Smythies. London, Hutchison and Co. Pp. 1–55.

Xenakis, Iannis

    1971. *Formalized Music.* Bloomington and London, Indiana University Press.

Young, R. W.

    1960. "Musical Acoustics." In *McGraw-Hill Encyclopedia of Science and Technology*, Vol. 8, p. 661.

Zimmer, Heinrich

    1956. *Philosophies of India.* New York, Meridian Books.

Index